Progress in Nonlinear Differential Equations and Their Applications
Volume 14

Editor
Haim Brezis
Université Pierre et Marie Curie
Paris
and
Rutgers University
New Brunswick, N.J.

Jean-Michel Morel
Sergio Solimini

Variational Methods
in Image Segmentation

with seven image processing experiments

Birkhäuser
Boston • Basel • Berlin

Jean Michel Morel
CEREMADE
Université Paris-Dauphine
Place de Lattre de Tassigny
75775 Paris Cedex 16, France

Sergio Solimini
Università degli Studi di Lecce
Srada provinciale Lecce-Arnesano
73100 Lecce Italy

Library of Congress Cataloging-in-Publication Data

Morel, Jean-Michel, 1953-
    Variational methods for image segmentation / Jean-Michel Morel,
Sergio Solimini.
        p.   cm. -- (progress in nonlinear differential equations and
their applications ; v. 14)
    Includes bibliographical references and index.
    ISBN 0-8176-3720-6   (acid-free)
    1. Image processing--Digital techniques--Mathematics.
2. Geometric measure theory.   I. Solimini, Sergio, 1956-
II. Title.   III. Series.
TA1637.M67   1994   94-36639
621.3'67--dc20   CIP

Printed on acid-free paper   *Birkhäuser* 🅱 ®

© 1995 Birkhäuser Boston

Softcover reprint of the hardcover 1st edition 1995

ISBN-13: 978-1-4684-0569-9   e-ISBN-13: 978-1-4684-0567-5
DOI: 10.1007/978-1-4684-0567-5

Typeset and reformatted from the author's disk in $\mathcal{AMS}$-T$_{\text{E}}$X

9 8 7 6 5 4 3 2 1

Contents

Preface

This book contains both a synthesis and mathematical analysis of a wide set of algorithms and theories whose aim is the automatic segmentation of digital images as well as the understanding of visual perception. A common formalism for these theories and algorithms is obtained in a variational form. Thank to this formalization, mathematical questions about the soundness of algorithms can be raised and answered.

Perception theory has to deal with the complex interaction between regions and "edges" (or boundaries) in an image: in the variational segmentation energies, "edge" terms compete with "region" terms in a way which is supposed to impose regularity on both regions and boundaries. This fact was an experimental guess in perception phenomenology and computer vision until it was proposed as a mathematical conjecture by Mumford and Shah.

The third part of the book presents a unified presentation of the evidences in favour of the conjecture. It is proved that the competition of one-dimensional and two-dimensional energy terms in a variational formulation cannot create fractal-like behaviour for the edges. The proof of regularity for the edges of a segmentation constantly involves concepts from geometric measure theory, which proves to be central in image processing theory. The second part of the book provides a fast and self-contained presentation of the classical theory of rectifiable sets (the "edges") and unrectifiable sets ("fractals"). This part contains a discussion of new uniform density properties of rectifiable sets, which prove extremely useful in image processing theory. Several image processing experiments and many figures illustrate algorithmic discussions and mathematical proofs.

This book will be accessible to graduate science students with some mathematical background. It will be of interest to mathematicians concerned with the interaction of analysis and geometry and to vision researchers.

Introduction

Natural, digital and perceptual images.

When one looks directly at scenes from the natural or the human world, or at any image (painting, photograph, drawing,...) representing such scenes, it is impossible to avoid seeing in them structures, which in many cases can be identified with real objects. These objects can be somehow concrete, as in photographs where we see trees, roads, windows, people, etc., or abstract perceptual structures, as the ones which appear in abstract paintings and can only be described in geometrical terms. However, we know that the "visual information" arriving at our retina, far from being structured, is purely local information for which a good model is given by digital images.

From the mathematical (and engineering) viewpoint, digital images simply are functions $g(x)$, where x is a point of the image domain Ω (the plane, a rectangle, ...) and $g(x)$ is a real number representing the "brightness" or "grey level" of the image at point x. This is the unstructured datum with which engineers have to deal in image analysis, robotics, etc.... And it also is somehow the basic datum which arrives at our retina. There is therefore not much in common between what we think we "immediately" see and what the physical information about light reflected by objects is. This striking difference between "images" in the engineering or the biological sense and "images" in the perceptual, artistic, semantic sense is well known, but the question is: How do we pass from the unstructured digital image to the structured perceptual one?

This question has been addressed in a scientific framework by psychologists of the Gestalt school in the twenties (Koehler [Koeh]) and then, with the first computers allowing to display synthetic images, by psychophysicists who did measurements of how much was perceived in the very first milliseconds after the arrival of an image to the retina (Julesz [Ju1, Ju2]). At the end of the sixties, the same problem was addressed in a very practical framework by engineers who wished to extract automatically useful information from digital pictures. The aim of this extraction was to build perception-driven robots and to better understand the human and animal vision. Then arose the *Segmentation Problem*, which is the object of this book.

The segmentation problem and its algorithms.

Segmenting a digital image means finding (by a numerical algorithm) its *homogeneous regions* and its *edges*, or *boundaries*. Of course, the homogeneous regions are supposed to correspond to meaningful parts of objects in the real world, and the edges to their apparent contours. Gestaltists and psychophysicists agree that such a segmentation process is at work at the very first stages of the visual perception process. In addition they proved that these first stages are highly independent of any learning or *a priori* knowledge of the world. In other words, the Vision research assumes that these first stages are accomplished by a *Geometry machine*, and it is also assumed that computer algorithms, working on digital images, should be able to do the same (Marr [Marr3], Koenderink [Koen2]). It is not the aim of this book to directly discuss this thesis, but rather to classify the proposed algorithms and their mathematical properties. More than a thousand algorithms have been proposed for segmenting images or detecting "edges". It is of course impossible (and unnecessary) to review them all, but the first part of this book (Chapters 1 to 4) is devoted to a classification of these algorithms and their translation from a discrete into a continuous framework (more adapted to the mathematical analysis). The result of this discussion was unexpected to the authors of this book because they became aware that under the very large diversity of numerical tools, **there essentially was only one segmentation (or "edge detection") model.** Indeed, as we hope these chapters will convince the reader, most segmentation algorithms try to minimize, by several very different procedures, one and the same *Segmentation energy.* This energy measures how smooth the regions are, how faithful the "analyzed image" to the original image and the obtained "edges" to the image discontinuities are.

One can write, and we indeed write in Chapter 4, the "most general" segmentation energy underlying all the analysed computational models. It has six or seven partly redundant terms, however, and therefore does not fit elegant mathematical analysis. If we keep the three more meaningful terms of the functional, we obtain the Mumford-Shah energy. Thus the Mumford-Shah variational model, although initially proposed as one model more, happens to somehow be **the general model** of image segmentation, and all the other ones are variants, or algorithms tending to minimize these variants. The Mumford-Shah model defines the segmentation problem as a joint smoothing/edge detection problem: given an image $g(x)$, one seeks simultaneously a "piecewise smoothed image" $u(x)$ with a set K of abrupt discontinuities, the "edges" of g. Then the "best" segmentation of a given image is obtained by minimizing the functional

$$E(u, K) = \int_{\Omega \setminus K} (|\nabla u(x)|^2 + (u - g)^2) dx + \text{length}(K).$$

The first term imposes that u is smooth outside the edges, the second that the piecewise smooth image $u(x)$ indeed approximates $g(x)$, and the third that the discontinuity set K has minimal length (and therefore in particular is as smooth as possible). The model is minimal in the sense that removing one of the above three terms would yield a trivial solution. Needless to say, such a simple functional cannot give a good account of the geometric intricacy of most natural images, or of our perception of them. What is expected from algorithms minimizing such a functional is a sketchy, cartoon-like version of the image, and these algorithms will give perceptually good results when the processed images somehow match this *a priori model*: contrasted images with objects presenting piecewise smooth surfaces. We shall see some experimental examples in Chapters 2 to 5.

The Mumford-Shah conjecture.

Two main problems have to be solved once we have fixed a single universal segmentation model. The first one is practical: How can we define algorithms minimizing the Mumford-Shah energy on computers? Now, as we pointed out, the algorithms have in many cases **preexisted** the theory and many of them, though perfectible, yield perceptually reasonable results, as we shall see in Chapters 2 to 5. The second question is simply a mathematical one: Is the model consistent? That is, do segmentations exist which indeed minimize the Mumford-Shah energy, are they unique and are the boundaries thus obtained smooth?

Mumford and Shah [MumS1] conjectured the existence of minimal segmentations made of a finite set of C^1 curves. So far, this conjecture has not been fully proved and only partial but meaningful enough results are at hand. The problem has proved difficult for the present mathematical technique because of the subtle interaction of the two-dimensional term (in u) and the one-dimensional term length(K). The same difficulty arises when one wishes to define a computer program minimizing the energy and we hope that the mathematical analysis will somehow clarify the numerical debate.

Edge sets and rectifiable sets.

As a matter of fact, the first mathematical task is to correctly define the functional $E(u, K)$. Indeed, we cannot *a priori* impose that an edge set K minimizing E is made of a finite set of curves. This is precisely what has to be proved, and if we imposed this condition to all "candidates" to be minimizers, we would have, for the time being, no existence theorem at all.

This kind of situation is classical in mathematical analysis and is dealt with by enlarging the Search Space, that is, in our case, by looking for a solution in a wider class of sets with finite length than just finite sets of curves. This is done by defining the "length" of K as its one-dimensional Hausdorff measure, which is the most natural way of extending the concept of length to non-smooth sets (Carathéodory, 1915 [Cara], Hausdorff, 1919 [Haus]). So we shall have to work in the main part of this book with "1-sets", i.e., sets with finite, positive, one-dimensional Hausdorff measure. The theory of these sets was developed by Besicovitch between 1928 and 1944 [Besi1, 2, 3], and was completed by Federer (1947, [Fed1]), Marstrand (1961 [Marst]), Mattila (1975 [Matt]) and Preiss (1987 [Prei]). As the dates indicate, it has not been a straightforward theory.

Now, it is a extremely suggestive theory for image processing. Besicovitch conjectured a remarkable structural classification of sets with finite m-dimensional Hausdorff measure. This conjecture was proved by himself and the above-mentioned authors. We shall now explain the results of the theory in the case $m = 1$, which is particularly significant for image processing, but they are identical in any dimension.

Besicovitch calls *rectifiable* any 1-set which is essentially contained in a countable family of curves with finite length. Clearly, the "edge sets" sought for in image processing belong to this class. A 1-set is called *fully unrectifiable* if it has no rectifiable part. Of course, any 1-set can be divided into its rectifiable and its unrectifiable part, but the question arises: How can we separate them? Here comes the remarkable Besicovitch discovery: Besicovitch conjectured and thereafter proved that there is a simple *local density criterion* to decide whether a point x of K belongs to the rectifiable part of K or not. This criterion is, from the computational viewpoint, clear cut: If x is in the rectifiable part, then the length of $K \cap B(x, r)$ is equivalent to $2r$ when r tends to zero. Otherwise, one can find balls $B(x, r)$ with arbitrarily small radii such that the length of $K \cap B(x, r)$ is less than $\frac{3}{4} 2r$. This is the *density criterion*. Another geometric criterion is the following. If x is in the rectifiable part, then K admits a tangent line at x, and if x is in the unrectifiable part, every cone with vertex x and arbitrarily small diameter contains parts of K with positive length. Here again, strong *local criteria* give an account of the (nonlocal) rectifiability property. Last but not least, the unrectifiable part has a "fractal" structure which, among other properties, implies that its projection on almost every line has zero length.

When a computer scientist has applied some *edge detector* to an image, he can see on the screen a set of points, and most edge detection or *segmentation devices* deal with so-called *connectivity algorithms*. What do these algorithms aim at? Clearly, a separation of the *rectifiable* part from the unrectifiable one. Indeed, "edges" are always assumed to be contained in curves or in finite unions of curves because they are assumed to correspond to the apparent contours of physical objects. Even if these objects may have "fractal" boundaries, like trees or clouds, the perception and the computer programs tend to define their boundary as a set of smooth lines. *The Besicovitch theory is a first answer as to the theoretical possibility of separating by local criteria the rectifiable part of a set from its unrectifiable part.* To our knowledge, no one of the Besicovitch criteria has been tried for "cleaning" edge sets from their spurious, unrectifiable part. The reason may be that the situation, in the numerical framework, is less contrasted: what really is aimed at is *a separation from a more rectifiable part from a less rectifiable part.*

We shall see that variational algorithms in image processing implicitly do this task, since minima of the Mumford-Shah functional are *uniformly rectifiable* in a sense which will be developed in the third part of this book. In any case, the second part of this book is devoted to a complete exposition of the Besicovitch theory. Our presentation does not follow the original Besicovitch presentation, which only works for dimension 1. (An excellent such presentation is done in Falconer [Fal1]). We tried to take into account all the new information about the geometric structure of rectifiable and unrectifiable sets given by the above mentioned Marstrand and Mattila contributions, as well as several new compactness results about sequences of m-sets. All of these results are proved because they are used in the third part of this book: We wished the exposition to be as complete and self-contained as possible.

What will be proved about the Mumford-Shah functional.

Although the Mumford and Shah conjecture is not yet proved, adequately weakened versions have been, and we shall list a series of questions which have been and will in this book be positively answered.

• Do minimal segmentations exist, and are they unique for each image? The answer is yes for existence and no for uniqueness, which matches the experimental intuition.

• Is the set of minimal segmentations small? Somehow, yes, because it is compact.

We now are concerned with the **structure** of the edge set.

• Are the edge sets obtained by the Mumford-Shah theory *rectifiable* in the Besicovitch sense? The answer is yes, and this is a first confirmation that the variational method is sound.

• Is the "edge set" made of a finite set of (not necessarily smooth) curves? The answer is unknown, but we shall see that the edge set is almost equal to a finite set of curves and can even be enclosed in a single curve with finite length, which is a remarkable result!
At this stage of the theory, one can deduce that the Mumford-Shah energy and conjecture are sound and that all the segmentation algorithms discussed in the first part of the book find, beyond their own experimental justification, a mathematical label of correctness.

To summarize, this book is divided in three parts:

• Modelization (Chapters 1 to 5),
• Geometric measure theory and the structure of sets with finite Hausdorff measure (Chapters 6 to 12), and
• Existence and structural properties of the minimal segmentations for the Mumford-Shah model (Chapters 13 to 16).

We tried to make this book as self-contained as possible. The first part only asks that the reader have some familiarity with the formalism of partial differential operators; no further mathematical knowledge is required. The second part is fully self-contained. It will, however, seem easier to readers having some knowledge of the Lebesgue measure and integration theory. The last part asks the reader to know the elementary distribution theory, but not more than the mere definitions of derivatives in the distributional sense. The few results about elliptic equations and variational problems used therein are either proved or explicitly admitted. The authors wish to thank Haim Brezis who had the idea that the present book could be written, and Ennio de Giorgi and Yves Meyer, who were a steady source of good suggestions and encouragement. The remarkable applications of the segmentation algorithm discussed in Chapter 5, developed at Cognitec-Inc by Lenny Rudin and Stanley Osher, brought surprise and excitement. We thank Isabeau Birindelli and Guy David for many useful comments. The first author thanks heartily his collaborators Luis Alvarez, Coloma Ballester, Vicent Caselles, Antonin Chambolle, Francine Catté, Tomeu Coll, Françoise Dibos, Manolo Gonzalez, Georges Koepfler, Frédéric Guichard and Christian Lopez, who kindly put to his disposition experimental results. Every successful image processing experiment condenses a lot of mathematical and algorithmic skill.

Part I

MODELIZATION

Chapter 1

EDGE DETECTION AND SEGMENTATION

In this chapter, we shall give a short survey of the segmentation problem as it appears in the research in Computer Vision and introduce the main ideas which lead to the choice of a variational approach. The terms like edge, segmentation, region *and* boundary *will not receive a single definition since several theories and algorithms for defining them will be compared. These terms will be therefore assumed, in the beginning of this chapter, in their heuristic meaning.*

1. Is there a general theory for segmentation?

This section is devoted to the relevance of the segmentation in Computer Vision and to some basic arguments in favour of a variational formulation. A good explanation of why the edge detection problem has become such an important and even obsessing problem can be found in Lowe [Low1]:

> *Edge detection plays an important role in many computer vision systems (and apparently in biological vision) by identifying points of intensity discontinuity in an image. The locations of these intensity discontinuities usually reflect underlying discontinuities in the geometry of surface reflectance of a scene and thereby discount the effects of varying illumination and imaging parameters. For this reason, edges are proved to be one of the most reliable low-level features for bridging the gap between image intensities and scene properties.*

A striking confirmation of the importance of edge detection in vision (which even leads the perceptual system to "see" illusory contours) was found by R. von der Heydt and his laboratory [HePB]. The responses of single cells in visual areas are codified by describing the visual field: the area within which moving or stationary bars and edges produces activity. They found, however, that many cells in visual area V2 responded when no actual stimulus was present in their visual field, but when edges outside this field produced illusory contours that cross this field.

Some of the earliest attempts to build computer vision systems limited their input data to TV-images of block-world scenes (e.g., Guzman [Guz]). Let us quote Zucker [Zu2]:

> *The belief was widespread that a suitable local edge operator could be found that would give a response which could be tracked into a*

line drawing of the scene. This line drawing would then serve as the low-level segmentation upon which subsequent processes would operate. Now, the problem with local edge operators is that they respond not only to the presence of edges but also to various noise configurations. This ambiguity led to substantial problems in the design of algorithms for joining these edge responses into the elusive perfect line drawing.

Early attempts to solve the edge-detection problem by global, close to variational, methods are Martelli [Mart] in 1972, Montaneri [Mont] in 1971, and Brice and Fennema [BriF] in 1970. As becomes clear from these papers, the "low-level" analysis of images already presented the two basic opposite requirements of the segmentation problem:

- One wishes to smooth the homogeneous regions of the picture with two scopes: noise elimination and image interpretation.
- On the other side, one wants to keep the accurate location of the boundaries of these regions. Such boundaries are called "step edges" (see [Marr2]).

This problem was raised more recently by Haralick and Shapiro [HarS]:

What should a good image segmentation be? Regions of an image segmentation should be uniform and homogeneous with respect to some characteristic such as gray tone or texture. Region interiors should be simple and without many small holes. Adjacent regions of a segmentation should have significantly different values with respect to the characteristic on which they are uniform. Boundaries of each segment should be simple, not ragged, and must be spatially accurate.

Achieving all these desired properties is difficult because strictly uniform and homogeneous regions are typically full of small holes and have ragged boundaries. Insisting that adjacent regions have large differences in values can cause regions to merge and boundaries to be lost.

Haralick and Shapiro conclude with the following assertion.

As there is no theory of clustering, there is no theory of image segmentation. Image segmentation techniques are basically ad hoc and differ precisely in the way they emphasize one or more of the desired properties and in the way they balance and compromise one desired property against another.

We cannot completely agree with this pessimistic conclusion. Indeed, the obvious fact that the segmentation criteria can vary a lot only indicates that a general theory of segmentation should not focus primarily on the details of these criteria. If we trust the other assertions of these authors, we shall be led to the following questions:

a) Have the segmentation criteria a generic form, which could be therefore studied in whole generality?

b) Is there a computational theory to automatically find a compromise between those competing criteria?

These questions might not have a good answer in the tremendously general framework of data analysis. However, the segmentation problem is much more specific (two-dimensional, geometrically invariant, ...). As we shall see, questions a) and b) can now be adressed in such a way that

• a common variational formulation can be given to most segmentation and edge detection problems; and

• most segmentation and edge-detection methods (and even the most heuristic) can be interpreted as attempts to solve this common variational formulation.

This point of view is certainly now well admitted in the vision community. It was implicit in Zucker's following remark [Zu2], stated many years ago:

> *The separation of edge and region-based approaches expresses more of a difference in initial orientation than a difference in practice.*

Now, what is a variational formulation? It consists of summarizing all criteria concerning a set of edges K in a single functional $E(K)$ with real value. In other terms, to any set of edges or "segmentation" K is associated a value $E(K)$ which states how "good" the segmentation is. A classical convention is to define E in such a way that $E(K)$ is better the lower it is. Is that a strong assumption to associate such a functional with a segmentation method? Certainly not. We shall give in this chapter four kinds of arguments in favour of an explicit variational formulation.

The first one is the complexity of the heuristic methods dealing with image segmentation. Let us take as a characteristic example Perkins' method for image segmentation [Per].

The nine successive steps of the proposed algorithm are:

A) Compute "edge regions" by a threshold on the gradient. An edge point is a point with high enough gradient.

B) Thin the gradient regions so that they become one-dimensional.

C) Expand the edges thus obtained into new edge regions in order to close the small gaps between edges.

D) Determine the different regions by a connectivity algorithm.

E) Eliminate the small regions.

F) Shrink again the edge regions.

G) Add new edge pixels to close the newly appearing gaps between regions.

H) Eliminate small edge regions.

I) Calculate properties of uniform intensity regions.

All of the properties aimed by the algorithm are certainly necessary for a "good" segmentation and are indeed sought by most theories of segmentation or edge detection. Other recent works on image segmentation have emphasized the need for the combination of numerous tools to obtain a sound segmentation (Gagalowicz and Monga [GagM], Pavlidis and Liow [PavL], Prager [Pra]). Now, the number of parameters of such methods might be reduced as much as possible. Only a rigorous axiomatic of the segmentation problem can provide this reduction.

The second argument in favour of a variational formalization of the segmentation problem is an *a priori* argument based on an obvious comparison principle. Any reasonable segmentation method must obey a comparison criterion. Indeed, given two segmentations, it must be able to state which is better. In other terms, a segmentation method implicitly orders all segmentations. Now, every order can be quantified by a functional $E(K)$ such that $E(K) > E(K')$ if K' is better than K, and it is no strong assumption that $E(K)$ should be a real number.

The third argument is factual (*a posteriori*). We shall verify in the next chapters that most practical segmentation methods lead to the minimization of an explicit (and simple) functional. Moreover, we shall explain in Chapter 4 a simple method for "translating" any heuristic method into a variational one and we shall apply it to several examples.

The fourth argument founding variational methods is axiomatic and consists of deducing the variational formulation from a classical axiomatization of image processing: the so-called "multiscale analysis". Multiscale analysis introduces in image processing a basic parameter, the "scale", which we shall denote by λ. There are of course several ways of defining the scale. Roughly speaking, the definition which we adopt relates it to the minimal size of the details which will be kept in the analysed image. Now, the precise meaning of the scale depends on the kind of multiscale analysis in consideration. Therefore, a common notation across several theories and variants corresponds to the fact that in the formalization to follow, the scale has the same role.

2. Principles of multiscale analysis.

The only basic parameter in computer vision algorithms is "scale" and, as we shall see on examples, computer vision algorithms are always stated with a "multiscale" formulation. We list in this section some elementary properties satisfied by most multiscale algorihms. We define a picture as a "grey level" or "brightness" real function $u_0(x)$ defined at each point ("pixel") of a domain Ω, which is generally the plane, a square or a rectangle.

Multiscale analysis is concerned with the generation, from a single initial picture u_0, of a sequence of simplified pictures $u_\lambda(x)$, where $u_\lambda(x)$ appears to be a rougher and sketchier version of u_0 as λ increases. In u_λ, details and features like edges are kept if their "scale" exceeds λ.

Denote by S_λ the map: $u_0 \mapsto S_\lambda(u_0) = u_\lambda$. In the image processing literature, $S_\lambda(u_0)$ is either a picture, denoted u_λ, or more generally a pair (K_λ, u_λ), where K_λ is a set of "boundaries" or "edges at scale λ".

The basic (and rather obvious) properties which any multiscale analysis must satisfy are:

1) Fidelity: $u_\lambda \to u_0$ as $\lambda \to 0$.

2) Causality: $S_\lambda(u_0)$ only depends on $S_{\lambda'}(u_0)$ if $\lambda > \lambda'$.

3) Euclidean invariance: If A is an isometry, then $S_\lambda(u_0 \circ A) = (S_\lambda(u_0)) \circ A$.

4) (In case of boundary detection) Strong causality: $K_\lambda \subset K_{\lambda'}$ if $\lambda > \lambda'$.

Since $S_\lambda(u_0)$ is obtained from u_0 by a "simplification", as axioms (1) and (2) imply, it is rather intuitive that this loss of information should be quantified by some energy functional $E(u_\lambda)$ such that $E(u_\lambda)$ decreases as λ increases. Moreover, the euclidean invariance of this functional, coming from axiom (3) makes it likely to be a Lebesgue or Hausdorff integral of some bidimensional and/or monodimensional terms (in case of boundary detection) depending on u_λ and K_λ. We shall verify that, in practical approaches, it is always the case. In the next three chapters, we examine a range of fundamental examples of multiscale analyses and discuss their variational formulation.

Chapter 2

LINEAR AND NONLINEAR
MULTISCALE FILTERING

In this chapter, we analyse the main theories of image multiscale filtering (and subsequent multiscale edge detection). The main point of the doctrine is that "edge detection" is a kind of differentiation; now, the image must be filtered (and smoothed) before differentiation. Thus "edge detection" becomes multiscale, like the filtering, and any edge detection theory is directly deduced from a multiscale filtering theory. We show how variational formulations are naturally associated with the main multiscale filtering theories.

1. The linear model (Hildreth, Marr, Witkin, Koenderink, ...).

As recalled by Zucker [Zu2], the edge detection theory of Hildreth and Marr is embedded in one of the most complete attempts to implement a general-purpose low-level vision system. Marr conjectured that the purpose of low-level vision is to construct a Primal Sketch, i.e., a symbolic description of the intensity changes in an image. His system begins by convolving local edge and bar masks over the image in several orientations and sizes. The responses of these masks are then parsed into a set of assertions dealing with the presence of various edge, line, and blob-like structures. Finally, a number of grouping operations attempt to eliminate noise responses and to cluster these initial symbols into assertions about longer lines and edges.

Let us develop with more details how edges are found, according to the original Hildreth and Marr theory [RosT],[Wi2],[YuP1]. The set of "edges" is assumed to be contained in the set of the points where the laplacian of the smoothed signal changes sign.

We shall follow the presentation and improvements of this theory made by Witkin [Wi2], Koenderink [Koen], and Canny [Can]. The low-pass filtering is made by convolution with gaussians of increasing variance. One easily understands the necessity of a previous low-pass filtering: if the signal is noisy, the gradient will have a lot of irrelevant maxima which must be eliminated. Of course, strong oscillations can be due to different causes, for instance, to the presence of "textures". Thus "edge detection" is associated with a family of integrodifferential operators.

In a recent paper, Torre and Poggio [TorP] compare the properties of several integrodifferential operators which have been proposed as edge detectors. One of the main results is a rigorous justification of *filtering before differentiation*. It justifies the use of suitable derivatives of Gaussian-like filters in edge detection. Indeed, they don't create edges as the accuracy of the filter decreases, a form of *strong causality*, and they ensure with high probability that zero-crossing edges form closed contours. (We shall give below a formal definition of "zero-crossing edges".)

Koenderink [Koen] noticed that the convolution of the signal with gaussians at each scale was equivalent to the solution of the heat equation with the signal as initial datum. Denoting by u_0 this datum, the "scale space" analysis associated with u_0 consists in solving the system

$$(2.1) \qquad \begin{cases} \dfrac{\partial u(x,t)}{\partial t} = \Delta u(x,t) \\ u(x,0) = u_0(x). \end{cases}$$

The solution of this equation for an initial datum with bounded quadratic norm is $u(x,t) = G_t \star u_0$, where $G_\sigma(x) = \frac{1}{4\pi\sigma} exp\left(-\frac{\|x\|^2}{4\sigma}\right)$ is the Gauss function.

The sequence $u_\lambda = u(\cdot, \lambda)$ clearly defines a multiscale analysis satisfying the properties 1 (fidelity), 2 (causality), and 3 (invariance) listed in section 1.2 of Chapter 1. Now, many attempts have been made to obtain or restore a strong causality property based on this multiscale analysis (Witkin [Wi2], Mallat and Zhong [MallZ]).

Restoration of the strong causality: Witkin's *scale space* theory.

Hildreth and Marr say that x is an edge point for the "scale" $\sqrt{t}$ if $\Delta u(x,t)$ changes sign and $|\nabla u(x,t)|$ is "large". Of course, this last condition introduces some *a priori* defined threshold. Unfortunately, it is enough to look at the "edges" found by this method [Marr3] to see that the edges at low scales give an inexact account of the boundaries which, according to our perception, should be considered as correct. (This is still true for the low-pass filtering of Canny [Can], which is generally used as the best linear filter for white noise elimination and edge detection). On the opposite side, if one makes a sharp low filtering, with small variance, all the edges will keep their correct location. Now, the "main" edges will be embedded in a crowd of "spurious" edges due to noise, texture, and so on. The "scale space" theory of Witkin [Wi2] proposes therefore to identify the main edges at a low scale, and then to "follow them backward" by making the scale decrease again. This method could theoretically give the exact location of all main edges. However, its implementation is rather heavy from the computational viewpoint and unstable because of the following of edges across scales and the multiple thresholdings involved in the edge detection at each scale.

Edge restoration by "inverting" the heat equation.

Osher and Rudin's theory [OshR] tries to deblurr images and to get as close as possible to the inverse heat equation by defining some conservative scheme

$$\frac{\partial u}{\partial t} = -|Du|F(\Delta u) \qquad \text{with initial value } u_0 ,$$

where F is a function such that $sF(s) \geq 0$. The big advantage of this method is to have a scheme which lets the image develop true edges, that is, shock lines along which $u(x,t)$ becomes discontinuous in x. This looks like the inverse heat equation because if, for example, $F(x) = x$ and if we get rid of the (positive) term $|Du|$, then the equation becomes the heat equation with time reversed.

The inverse problem: from the edges to the picture.

Another way of playing with the edges is proposed by S.G. Mallat and S. Zhong [MallZ], who follow David Marr's suggestion that the Hildreth-Marr edge representation at several dyadic scales should be a complete representation of a picture. They propose an algorithm, based on some stability properties of the wavelet transform, which closely reconstructs the picture from the edges at five scales. An interesting and suggestive fact of this algorithm is that if one removes the "spurious edges" (i.e. the edges which have low gradient or cannot be prolongated), then the reconstruction is still almost perfect. **This justifies** *a posteriori* **the importance given to edge-detection and enhancement in the literature**. An important application of this algorithm is selective smoothing. Indeed, these authors are able to clean up the picture by eliminating the "spurious" edges (i.e., the edges which are not stable across several scales) and thereafter reconstructing the picture. Another work worth mentioning on the same direction is by Hummel and Moniot [HumM], in which another algorithm for image reconstruction from edges is proposed. They give a mathematical proof of this reconstruction property, but only for continuous-scale edges. Their proof generalizes Logan's theorem, which was proved for one-dimensional signals.

2.2 The Perona and Malik theory

Variational formulation of the Hildreth-Marr theory.

As we announced in the preceding chapter, there is an energy naturally associated with the Hildreth-Marr multiscale analysis. It gives an evaluation of the amount of information contained in u by

$$E(u) = \int_\Omega |\nabla u|^2 \, dx \, .$$

This functional is the right one to be associated with model (2.1) *because the heat equation is the natural descent method associated with this energy.* If one looks at a discretization of the evolution equation, it can also take a variational form, since it is $u_t - u_0 = t\Delta u$, which is the solution of a variational problem. Roughly speaking, numerically solving the heat equation is therefore equivalent to minimizing the following functional iteratively, with $t = \lambda^2$ small:

$$E_\lambda(u) = \lambda^2 \int_\Omega |\nabla u|^2 \, dx + \int_\Omega (u - u_0)^2 \, dx \, .$$

This formulation has its own interest: it makes the scale parameter appear explicitly as the multiplier of the smoothing term. The solution u_λ of this problem can be defined as the "analysis of u at scale λ" as well as the solution of the heat equation $u(x, \lambda^2)$. Notice, however, that the solutions v_λ of the minimizing problem associated with E_λ do not satisfy the causality property.

2. The Perona and Malik theory.

The above-mentioned attempts to restore the *strong causality* are quite close to an important improvement and generalization of the edge detection theory proposed by Perona and Malik [MaliP1]. Their main idea is to introduce a part of the edge detection step into the filtering itself, allowing an interaction between scales from the beginning of the algorithm. They propose to replace the heat equation by a nonlinear equation:

$$(2.2) \qquad \begin{cases} \dfrac{\partial u}{\partial t} = \operatorname{div}\left(f(|\nabla u|)\nabla u\right) \\ u(0) = u_0. \end{cases}$$

In this equation, f is a smooth nonincreasing function with $f(0) = 1$, $f(s) \geq 0$, and $f(s)$ tending to zero at infinity. The idea is that the smoothing process obtained by the equation is conditional: If $\nabla u(x)$ is large, then the diffusion will be low and therefore the exact location of the edges will be kept. If $\nabla u(x)$ is small, then the diffusion will tend to smooth still more around x. Thus the choice of f corresponds to a sort of thresholding which has to be compared to the thresholding of $|\nabla u|$ used in the final step of the classical theory. Since this thresholding introduced a nonlinear device anyway, it was natural to use it earlier in the method, in the smoothing process itself. Notice that the new model reduces to the linear model of heat equation when $f(s) = 1$. The experimental results obtained by Perona and Malik are perceptually impressive and show that an edge detector based on this theory gives edges which remain much more stable across the scales, *making therefore the backward following of edges across scales unnecessary.* (This "strong causality" property has been sought after by most vision researchers in the two last decades and is much better attained by the Perona and Malik model than by the linear model.)

Variational formulation of the Perona and Malik model.

As in the preceding section let us look for the energy associated with the Perona and Malik model. Set $\tilde{f}(a) = f(\sqrt{a})$. Equation (2.2) can be written as

$$\begin{cases} \dfrac{\partial u}{\partial t} = \text{div}\,(\tilde{f}(|\nabla u|^2)\nabla u) \\ u(0) = u_0. \end{cases}$$

Now set $\tilde{F}(t) = \int_0^t \tilde{f}(s)\,ds$ and define an energy E by

$$E(u) = \frac{1}{2}\int_\Omega \tilde{F}(|\nabla u|^2)\,dx \;.$$

Suppose that u is a minimum of E. Calculating the first variation of E for every test function φ yields

$$\frac{E(u+t\varphi) - E(u)}{t} \approx \int_\Omega \tilde{f}(|\nabla u|^2)\nabla u \nabla \varphi\,dx$$
$$= -\int_\Omega \varphi \,\text{div}\left(\tilde{f}(|\nabla u|^2)\nabla u\right)\,dx$$
$$= -\int_\Omega \varphi \,\text{div}\,(f(|\nabla u|)\nabla u)\,dx.$$

So equation (2.2) is a steepest-descent method for the energy defined above and we can associate with it, as we did for the Marr-Hildreth-Witkin-Koenderink theory, a variational multiscale analysis

$$E_\lambda(u) = \lambda^2\int_\Omega \tilde{F}(|\nabla u|^2)\,dx + \int_\Omega (u - u_0)^2\,dx.$$

This last variational formulation has been proposed by Nordström [Nor]. He introduces a new term in the equation (2.2) which forces $u(x,t)$ to remain close to u_0. Because of the forcing term $u - u_0$, the new equation

$$\frac{\partial u}{\partial t} - \text{div}\,(f(\nabla u)\nabla u) \;=\; u_0 - u$$

has the advantage of having a nontrivial steady state, eliminating therefore the problem of choosing a stopping time.

Discussion.

A confirmation of the soundness of Perona and Malik's approach is given in Nitzberg and Shiota [NiS]. Their method is related to adaptive filtering methods which where introduced for TV images denoising by Graham [Gra]. The idea is to blur selectively and anisotropically the signal with "oblong" gaussians. The gaussian used for blurring at a point x depends on the intensity and direction of the gradient in the neighbourhood. Roughly speaking, the blurring will be faster in the direction orthogonal to the gradient. Therefore, the signal will be smoothed on both sides of an edge, but the edge is conserved. Since a corner is the crossing point of two edges, there will be two directions of "nondiffusion" instead of one, and therefore corners are well conserved by this method. Nitzberg and Shiota prove by a scaling argument that as the size of the Gaussians tends to zero, their diffusion method tends to a partial differential equation analogous to the Perona and Malik model. Thus the Nitzberg and Shiota method is a kind of "interpolation" between the classical linear theory and the Perona and Malik model, and conversely, the Perona and Malik model can be interpreted as an adaptive smoothing method.

However, the Perona and Malik model has several serious practical and theoretical difficulties. The first difficulty is a straightforward objection that the authors acknowledge. Assume that the signal is noisy, with white noise for instance. Then the noise introduces very large, in theory unbounded, oscillations of the gradient ∇u. Thus, the conditional smoothing introduced by the model will not give good results, since all these noise edges will be kept!

The second difficulty arises from the equation itself: among the functions f which Perona and Malik consider advisable, one finds functions of the type $f(s) = e^{-s}$ or $f(s) = (1+s^2)^{-1}$ for which no correct theory of equation (2.2) is available. Indeed, in order to obtain both existence and uniqueness of the solutions, the function $sf(s)$ must be nondecreasing. If this condition is not satisfied, one can observe for some functions f with

$sf(s)$ nonincreasing a nondeterministic and therefore unstable process; the same picture can in theory be the initial condition of solutions divergent in time (see [HolN] for simple and explicit examples). In practice, that means that very close pictures could produce divergent solutions and therefore different edges. However, it is reasonable to try to define a model where the function f, which is a sort of thresholding, could be quickly decreasing, as e^{-s} for instance.

A synthesis of the Marr-Hildreth and Perona-Malik models.

In [CatCLM], a synthesis of Perona and Malik's ideas is proposed that avoids the above-mentioned difficulties: it is robust in the presence of noise and consistent from the formal viewpoint mentioned above.

Define the "selective smoothing of u_0 at scale $\sqrt{t}$ based on estimates at the scale σ" as the function $u(x,t)$ satisfying

(2.3)
$$\begin{cases} \dfrac{\partial u}{\partial t} = \text{div}\left(f(|\nabla G_\sigma \star u|)\nabla u\right) \\ u(0) = u_0 \end{cases}$$

where $G_\sigma(x) = \frac{1}{C\sigma}\exp(-\frac{\|x\|^2}{4\sigma})$.

As we have seen, $G(x,t) = G_t(x)$ is nothing but the fundamental solution of the heat equation. Therefore the term $(\nabla G_\sigma * u)(x,t)$ which appears inside the divergence term of (2.3) is simply the gradient of the solution at time σ of the heat equation with $u(x,t)$ as initial datum. Therefore, it appears to be an estimate of the gradient of u at point x, obtained by the classical Marr-Hildreth-Witkin theory recalled above. Thus, the modification of the model of Perona and Malik consists in replacing the gradient $|\nabla u|$ by its estimate $|\nabla G_\sigma * u|$. As it is proved in [CatCLM], this slight change of the model is enough to avoid both above-mentioned inconsistencies of the Perona and Malik model. The equation diffuses at a point with more or less strength, according to the estimate of the gradient. This estimate is achieved by the new term. This term, like in Witkin's theory, serves to recognize the location of the main edges. This information is then used in the equation to avoid too much diffusion at these locations.

However, this last model still keeps some of the drawbacks of the previous models. First it has no clear geometric interpretation, because the term inside the divergence is hybrid and combines the estimate of the gradient and the gradient. Moreover, even if existence and uniqueness are proved, the stability of the model as the scale parameter σ tends to zero is generally not true, because the limit model can "invert" locally the heat equation. As we shall see, these drawbacks are related to the excessive generality of the Perona and Malik model.

The meaning of "scale" in nonlinear models.

As commented in [CatCLM], the function G to be considered can be any "low pass filter", or, to use the calculus terminology, any smoothing kernel. However, in order to preserve the notion of scale in the gradient estimate, it is convenient that this kernel depends on a scale parameter. A good and classical example is, as mentioned above, the gaussian. It is important to keep this particular case in mind. Indeed, a question which arises immediately in the consideration of model (2.3) is what time is best for "stopping" the evolution of the signal $u(x, t)$. Now, one may appeal to the Witkin model to answer this question: according to this model, time is interpreted as a "scale factor". (More precisely, the solution $u(x, t)$ at time t corresponds to a spatial scale $\sqrt{t}$. Indeed, roughly speaking, $u(x, t)$ appears in the Witkin model as a smoothed version of u_0 obtained by convolving it with a filter of spatial width $\sqrt{t}$.) Thus in the model (2.3) it is coherent to correlate the stopping time t and the time introduced via the estimator G_σ. One should therefore choose a stopping time t of the order of σ. Then the scale above which the signal is smoothed in regular zones of the image will be of the order of $\sqrt{t}$.

On the parts of the picture where edges are present, the situation is different. Since the scope of the equation is to delay diffusion in these regions, the scale at which edge information is lost will depend on the shape of the thresholding function f. Therefore, even if we wrote above that one should stop the equation at a time of the order of σ, this is rather a lower estimate and there is no inconsistency to look at what happens to the signal $u(x, t)$ for the time parameter larger than σ.

3. Anisotropic diffusion and "mean curvature motion".

The intuitive idea of "edges" is that they are piecewise smooth and have a tangent at "most points". Therefore, it seems natural to modify the diffusion operator so that it diffuses more in the direction parallel to the edge and less in the perpendicular one. The Perona and Malik theory can be interpreted as a way to do such an "anisotropic diffusion". In the extreme case, one could think of a diffusion which is made only in the direction of the edge: such a diffusion would keep exactly the location and sharpness of the edge, while smoothing the picture on both sides on this edge.

Let us first consider this limit case in a linear framework. It is not difficult to see that the diffusion equation which does not diffuse at all in the direction of the gradient ∇u can be written as

(2.4)
$$\frac{\partial u}{\partial t} = \Delta u - \frac{D^2 u(\nabla u, \nabla u)}{|\nabla u|^2} .$$

The first term, the Laplacian, is the same as in Scale Space theory, and the second term is an "inhibition" of the diffusion in the direction of the gradient. Let us denote by ξ the coordinate associated with the direction orthogonal to $|\nabla u|$. Therefore a formulation of the preceding equation with respect to this new coordinate is

$$\frac{\partial u}{\partial t} = u_{\xi\xi} ,$$

where, of course, ξ depends on ∇u.

In a quasi "divergence form", the equation can also be written as

$$\frac{\partial u}{\partial t} = |\nabla u| \, \text{div} \left(\frac{\nabla u}{|\nabla u|} \right)$$

and in a more literal formulation (denoting by (x, y) the cartesian coordinates of x),

$$\frac{\partial u}{\partial t} = \frac{u_y^2 u_{xx} - 2 u_x u_y u_{xy} + u_x^2 u_{yy}}{u_x^2 + u_y^2} .$$

This equation, which models the "mean curvature motion", has recently received a lot of attention because of its geometrical interpretation: indeed, at least formally (see S. Osher and J. Sethian [OshS], L.C. Evans and J. Spruck [EvS]) the level sets of the solution move in the normal direction with a speed proportional to their mean curvature.

A theory of weak solutions of this equation, based upon the "viscosity solution" theory, has been proposed by Y. G. Chen, Y. Giga and S. Goto [ChGG]; Y. Giga, S. Goto, H. Ishii and M.H. Sato [GiGIS]; and H.M. Soner [Son].

4. Image restoration by mean curvature motion.

Let us now see which equation is given by the combination of the degenerate diffusion with the above-mentioned improvements of the Perona and Malik theory: Alvarez, Lions and Morel [AlvLM] propose and study a class of nonlinear parabolic differential equations for image processing based on the above-explained "mean curvature motion"

(2.5)
$$\begin{cases} \dfrac{\partial u}{\partial t} = f(|\nabla G_\sigma \star u|) \, |\nabla u| \, \text{div} \left(\dfrac{\nabla u}{|\nabla u|} \right) \\ u(x, 0) = u_0(x). \end{cases}$$

Roughly speaking, the interpretation of the terms of the equation are as follows:

a) The term $|\nabla u|$ div $\left(\frac{\nabla u}{|\nabla u|}\right) = \Delta u - \frac{D^2 u(\nabla u, \nabla u)}{|\nabla u|^2}$ represents a degenerate diffusion term, which diffuses u in the direction orthogonal to its gradient ∇u and does not diffuse at all in the direction of ∇u. The aim of this term is to make u smooth on both sides of an "edge" with a minimal smoothing of the edge itself. (An edge is defined as a line along which the gradient is "large".)

b) The term $f(|G \star \nabla u|)$ is used for the "enhancement" of the edges. Indeed, it controls the speed of the diffusion: If ∇u has a small mean in a neighbourhood of a point x, then this point is considered as the interior point of a smooth region of the image and the diffusion is therefore strong. If ∇u has a large mean value on the neighbourhood of x, this point is considered as an edge point and the diffusion spead is lowered, since $f(s)$ is small for large s.

Thus, the proposed model is a selective smoothing of the image, where the "edges" are relatively enhanced. It preserves the strong causality as much as possible.

Like the preceding models, this model depends upon two parameters:

- A "contrast" function, represented by f, decides whether a detail is sharp enough to be kept. This parameter corresponds to the final threshold on $|\nabla u|$ in the Hildreth-Marr theory.

- A "scale parameter", given by the variance of G, fixes the minimal size of the kept details in the processed picture.

Variational interpretation of the "mean curvature motion".
The Osher and Rudin variational model.

If $f(s) = \frac{1}{s}$ and $\sigma = 0$, we obtain as a limit case of this new model the equation

$$(2.6) \qquad \frac{\partial u}{\partial t} = \text{div} \left(\frac{\nabla u}{|\nabla u|}\right)$$

which is a particular case of the Perona and Malik model. Equation (2.6) corresponds to the descent method $(\partial u / \partial t = -\nabla E(u))$ associated with the energy functional

$$E(u) = \int_\Omega |\nabla u(x)|\, dx.$$

As we did for the Hildreth-Marr theory, we can associate with this evolution equation a multiscale energy. This energy has to be minimized when the evolution equation is discretized in time:

$$E_\lambda(u) = \lambda^2 \int_\Omega |\nabla u|\, dx + \int_\Omega (u - u_0)^2\, dx \ .$$

How can we interpret this energy? A particular case for u, which also is the ideal case for the detection of "step edges", is when $u_0 = \chi_A$ is the characteristic function of some set A with smooth boundary. Then $E(u_0)$ is nothing but the perimeter of A, and the evolution equation can therefore be interpreted as a smoothing of the boundary of this characteristic function which tends to preserve the "edges". This variational model and its variants has been proposed and developed by Osher and Rudin [OshR] for denoising images.

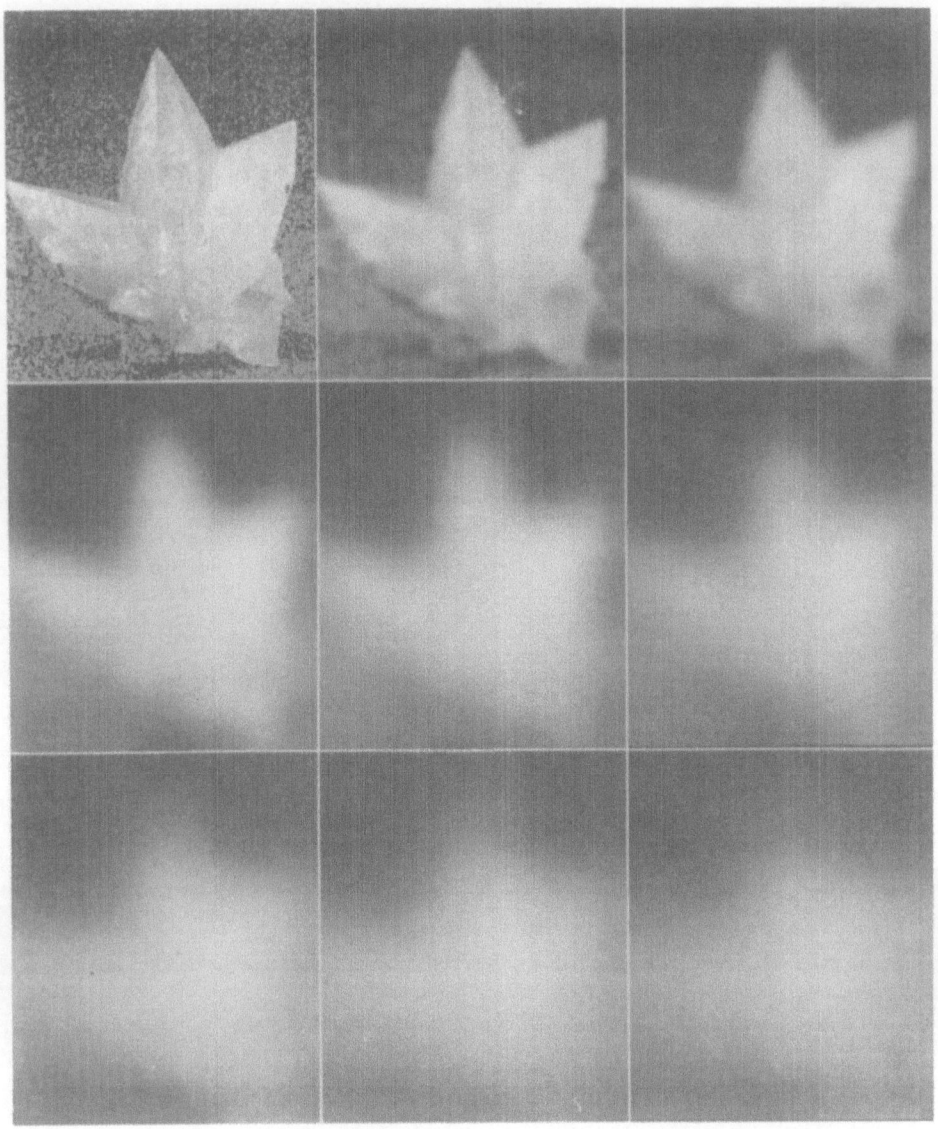

Heat equation as a multiscale analysis　　　　　　　**Author: L.Alvarez.**

The heat equation $\frac{\partial u}{\partial t} = \Delta u$ is applied to an initial image $u(0)$ which represents the picture of a calcite stone. We display, left to right and down, the result of the analysis for the scales: $\frac{t^2}{2} = 0, 2, 4, 10, 13, 16, 20, 23, 26.$

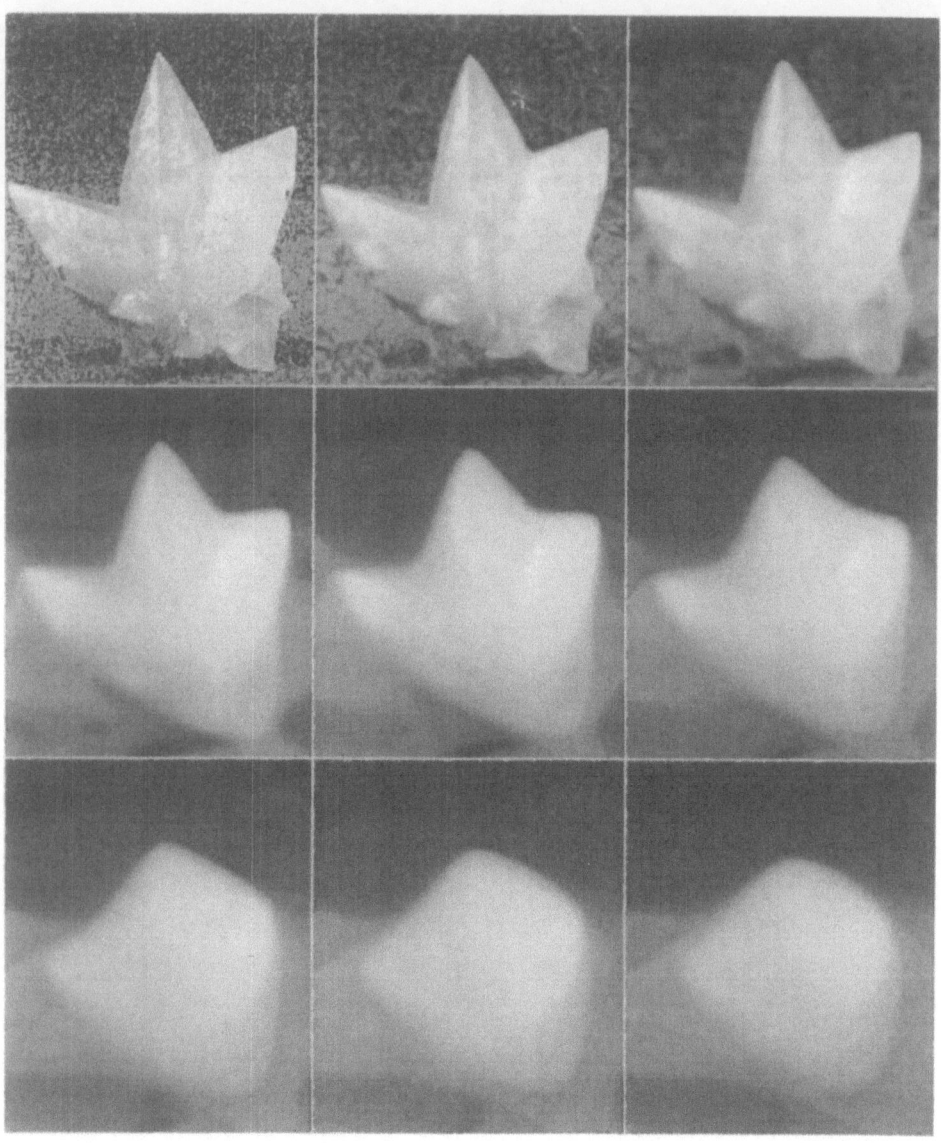

Affine invariant curvature motion [AGLM1,2]   **Author: L.Alvarez.**

In this picture we present the evolution accross the scales using the affine invariant curvature motion $\frac{\partial u}{\partial t} = |Du|(t div \frac{Du}{|Du|})^{\frac{1}{3}}$ of an initial image $u(0)$ which represents the picture of a calcite stone. We display, left to right and down, the result of the analysis for the scales: $t = 0, 2, 4, 10, 13, 16, 20, 23, 26.$

Chapter 3

REGION AND EDGE GROWING

The general idea of the region growing methods is to create a partition of the picture into homogeneous regions by any kind of process which starts with small regions and then grows them according to some homogeneity criterion. The general idea of edge growing methods is to create an initial set of ("fine scale") edges (e.g., by one of the above-mentioned theories) and to connect the edges iteratively, according to some criterion based on proximity and orientation. Both methods are "dual" and can be made to interact in "hybrid growing methods". In this chapter, we shall review and formalize the three above-mentioned kinds of growing methods and we shall, as in the last section, associate with them a variational formulation. This formulation is obtained by identifying for each method an energy for which the growing process appears as a steepest descent algorithm.

1. Formalization of region growing methods (Brice and Fennema, Pavlidis, ...).

One of the first attempts to do segmentation by region growing is the "regional neighbour search" of Muerle and Allen ([MueA], 1968). We shall adopt the terms of a classification of these methods later proposed by Zucker [Zu1] and Rosenfeld and Kak [RosK]. They distinguish between "region aggregation" and "region growing":

- Region aggregation (which is the properly called "region growing") begins with a "seed region" and makes the region grow by aggregation of pixels.

- Region growing: *"Start with a set of highly uniform regions, e.g., connected components of constant grey level, or, more generally, regions extracted by some preliminary segmentation process. Growth then takes place by starting with one of these regions and merging neighbouring regions with it, one at a time"* ([RosK, p. 141]). It is important to notice that the regions grown from seed regions may overlap, in contrast with the definition of segmentation as a partition given by Horowitz and Pavlidis (see below).

- Partitioning: *"Many of the methods of region growing can also be used to partition the picture into regions. More generally, picture partitions can be constructed by starting with an initial partition*

*and allowing both merging and splitting of regions. The initial
partition may be trivial (e.g., the entire picture is the sole region;
each pixel or $k \times k$ block of pixels is a region; the connected components of constant grey level are the regions; and so on), or it can
be the result of a previous segmentation process"* ([RosK, p. 145]).

Pavlidis [Pav1] is the first author to notice that the segmentation problem implies the introduction of a smooth approximate version of the image
on each region. The main point of his approach is to define a family $\mathcal{F}$
of approximating functions and to try to find regions O_i and functions u_i
of the family for which $\|u_i - u_0\|$ is small enough, where the norm is an
appropriate one for functions defined on O_i. If, for instance, the family
$\mathcal{F}$ is the family of locally constant functions, then the regions which are
sought are regions where u_0 is as constant as possible. The number of
regions will be decreased by the merging of regions whose constants are
close. Horowitz and Pavlidis [HoroP2] probably proposed the first general definition of region growing, which was generalized into a "split and
merge" paradigm, with an approximation variational criterion:

*Take an arbitrary initial partition in pyramidal data structure;
split regions with large approximation errors; merge adjacent regions
with similar approximation; and group similar squares into irregular
regions.*

*The notion of splitting is now algorithmically joined with the orig-
inal merging approach. The overall orientation is a functional approximation in which regions are described in terms of an approximating function. The closedness of the approximation is evaluated
by an error norm.*

The "split and merge principle", developed by Pavlidis and Horowitz,
is to merge adjacent regions having similar approximations and to split
those regions that have large error norms. One of the possible strategies
is to split the whole picture into four nonoverlapping squares, and to
split again those whose dissimilarity measure is too high, and so on. This
induces a so-called "pyramidal structure" on the picture, where at the
bottom are the single pixels as elementary regions, and at the top is the
whole picture. At each intermediate level, the picture is divided into
squares of the same size and each level has four times fewer squares than
the preceding. Thus the "split and merge" consists in moving up and
down in this pyramid, according to the merging or splitting operations,
thus decreasing the approximation norm, until a minimum is attained.
Notice that the height in the pyramid is nothing but a scale parameter.

The final result of a region growing method is always a partition of the image, but the "growing" presentation already implies the use of a scale parameter. Typically, this parameter is initially very small and increases during the region growing process. Thus all the methods quoted above are multiscale and satisfy the principles 1, 3 and 4 stated in Chapter 1. They can all be summarized (or improved) in the following generic algorithm:

Generic multiscale region growing algorithm.

Fix an initial scale λ_0 very small. Start with an initial partition of pixels or very homogeneous regions.

a) *Merge all pairs of regions whose merging "improves" the segmentation. Increase the scale parameter λ. Iterate.*

b) *If convenient, decrease λ and split the regions whose splitting might improve the segmentation. Go to step (a).*

In many methods, the scale parameter is not explicit but rather appears as a threshold parameter which evolves during the region growing process. Now, as in the methods that we presented in the preceding chapters, one can distinguish between the multiscale presentations and the energetic presentation. For instance, Horowitz and Pavlidis directly define a segmentation as an optimal partition:

If $u_0(x)$ is the brightness function of a picture defined on a domain D, then we may attempt to divide D into the minimum number of regions where $u_0(x)$ satisfies certain constraints (e.g., is approximately constant).

Of course, this definition depends on a hidden scale parameter, which is implicit in the constraints.

Region growing with overlapping regions.

In contrast with pure partitioning algorithms based on "split and merge", the region growing methods allow regions to overlap. For instance A. Leonardis, A. Gupta and R. Bajcsy [LeoGB] present an algorithm aimed at dividing the image into regions well described by a polynomial spline. The ideas of this paper are quite similar to P.J. Besl and R. Jain [BeslJ] but go further in the analysis of the model selection for the overlapping regions. The first step of the algorithm is a region growing. From seed regions located on a grid-like pattern, regions are grown. The growing of a region stops when the quadratic distance between the polynomial model and the image becomes greater than some *a priori* fixed threshold. The growing procedure outputs many different regions of which *many are partially or totally overlapped*. The criteria for selecting the models are first that their number is as small as possible, second that the size of each model is as large as possible, and third that the error measure between the original data and the recovered models is small. Another region growing method with overlapping, the Nitzberg and Mumford "2.1 sketch", will be discussed later on.

Variational formulation of the region growing methods.

We shall follow the remark made in Section 1.2 of Chapter 1, according to which there is a "natural" energy associated with any multiscale analysis: This energy measures the amount of information contained in the smoothed version of u_0. In fact, a region growing method provides two results at each scale: the boundaries of the obtained regions, whose union is denoted by K_λ, and the smoothed version u_λ of u in each region. The amount of information remaining in the segmentation has therefore two terms: one measuring the amount of boundaries K_λ and their smoothness, and one measuring the smoothness of u_λ on each region.

The simplest energy associated with a segmentation (u, K) obtained by region growing therefore is

$$E(u, K) = \mu \text{ length} (K) + \int_{\Omega \setminus K} |\nabla u|^2 dx .$$

If we follow the formal approach developed in Chapter 2, we can associate with the multiscale region growing scheme a discretized scheme where the scale appears explicitly. We are led to define a variational multiscale energy by

$$E(u, K) = \lambda^2 \int_{\Omega \setminus K} |\nabla u|^2 \, dx + \mu \lambda^2 \int_K d\sigma + \int_{\Omega \setminus K} (u - u_0)^2 \, dx ,$$

where $d\sigma$ denotes the one-dimensional Hausdorff measure (or "length" measure) restricted to K. So the second integral simply is the length of K. This energy is known as the Mumford and Shah functional (see [MumS1],[MumS2]). When u is imposed to be constant on each region, which is a reinforcement of the smoothness condition imposed by the first term, we are led to the functional

$$E(u, K) = \lambda^2 \int_K d\sigma + \int_{\Omega \setminus K} (u - u_0)^2 \, dx .$$

As we shall see in the next chapter, this functional is the simplest possible one among those associated with the region growing methods. The Mumford and Shah formulation is a simple instance of a general variational formulation of image processing problems due to Terzopoulos [Te3]. Following ideas of Tikhonov and Arsenin [TiA] for the solution of "ill-posed" problems, this author has proposed a generic variational formulation for most vision problems. According to Terzopoulos, the energy associated with each vision problem has two terms in competition: a "regularity" term and an approximation term. The first term penalizes unsmooth estimates of the picture and the second estimates those which are not close enough to the picture. (The same idea has been developed by Kass, Witkin and Terzopoulos [KasWT] for global edge detection and we shall discuss it in the next section.)

2. Edge growing methods.

A good justification of multiscale edge growing can be found in Lowe [Low2]:

> *While edge detection is an important first step for many vision systems, the linked lists of edge points produced by most existing edge detectors lack the higher level of curve description needed for many visual tasks. For example, they do not specify the tangent direction or curvature of an edge or the location of tangent discontinuities... The multiscale analysis of curves is complementary to any multiscale detection of the original edge points.*

The most obvious method for transforming local edge information into a global one is the "edge linking" described in Rosenfeld and Kak [RosK, Vol.2]:

> *When large pieces of edges or curves have been extracted from a picture, they can be further linked into global curves. The criteria for linking two curves into a large curve might depend on their strength (i.e., the average strengths of the original responses), lengths, and mutual alignment or "good continuation". Of course, if global curves of specific shapes are desired, this can also be taken into account in defining the linking criteria.*

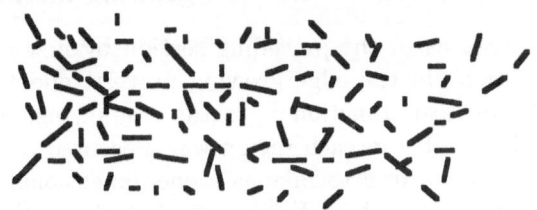

Figure 3.1 : Edge map with two "broken edges". How to find them by algorithm?

Another good description of edge linking, quoted below, is given by Weiss and Boldt [WeiB], who base their edge growing method on Hildreth-Marr's edge detection technique (see also Bajcsy and Tavakoli [BajT]).

> *Zero-crossing points of the Laplacian of intensity images and the gradients at those points are used to extract initial information about local edges. These edges are input to a <u>hierarchical</u> linking and merging algorithm. Edges are linked on both geometric and intrinsic properties, e.g., if their endpoints are close and their contrasts are similar. In the merging process, a sequence of linked edges is replaced by a longer straight line if the approximation is sufficiently good.*

We underlined "hierarchical" in the quotation because this term means once again that edge linking methods are multiscale. As in the preceding section on region growing, we now give a generic definition of edge growing and a formula for the natural associated energy.

The initial datum for this algorithm is a set of edges K_0. Each edge e of K_0 is defined as a curve with a length and a "strength" which is the mean value of $|\nabla u_0|$, or derivative of u in the direction normal to the edge, $\left|\frac{\partial u_0}{\partial n}\right|$. If the edge is smooth enough, its curvature may be also considered.

Generic multiscale edge linking algorithm.

a) *Begin with a set K_0 of "fine scale" edges and λ very small.*

b) *Link every pair of edges for which the linking is convenient for the scale depending criterion and smooth the linked curve. We obtain K_λ.*

c) *Increase λ and go back to (b).*

Variational formulation for the edge growing methods.

We shall proceed as in the preceding section by first identifying the "loss of information" in the edge growing process, and then by giving the associated multiscale functional. In the case of region growing, the increasing regularity of u in each region was measured by some integral of its gradient. It is natural to take as a one-dimensional equivalent an integral depending on length and curvature (which are the lowest order geometrically intrinsic measures on a curve). The loss of information in the region growing was also given by the length of the boundaries which were removed. Our one-dimensional equivalent is of course the cardinality

of the tips of the curves (the zero-dimensional measure of ∂K_λ). Therefore, the "natural" functional associated with the edge growing methods
is

$$E(K_\lambda) = \int_{K_\lambda} \left(1 + \text{curv}(\sigma)^2\right) d\sigma + \mu \, \text{Card}(\partial K_\lambda) \,.$$

This is the "smoothing" term of the method. By analogy with the other
multiscale methods, we need to define a simple "fidelity" term which
measures how good the approximation of K_0 by K_λ is. The simplest
term of the kind is length $(K_\lambda \setminus K_0)$. However, if the method tends to
smooth the edges, this term is no good measure of the proximity of K_λ to
K_0 because K_λ will move by the smoothing operation. Therefore, we can
simply use the same edge criterion for K_λ as the one which was used for
K_0. Since K_0 is found by an edge detector, which consists of detecting
maxima of the gradient of u_0, we can simply measure the fidelity of K_λ
to edges by $-\int_{K_\lambda} |\nabla u(\sigma)|^2 \, d\sigma$, or, more accurately, by $-\int_{K_\lambda} |\frac{\partial u}{\partial n}(\sigma)| \, d\sigma$,
which measures the "strength" of the edge.

Thus the most simple and generic multiscale variational energy associated with edge growing methods is (we omit the obvious normalization
constants and set all nonlinearities to be quadratic for simplicity):

$$E_\lambda(K_\lambda) = \lambda \left(\int_{K_\lambda} \left(1 + \text{curv}(\sigma)^2\right) d\sigma + \mu \, \text{Card}(\partial K_\lambda) \right)$$
$$- \int_{K_\lambda} |\nabla u(\sigma)|^2 \, d\sigma - \int_{K_\lambda} \left| \frac{\partial u}{\partial n}(\sigma) \right| d\sigma + \text{length}(K_\lambda \setminus K_0).$$

Not all terms in the preceding functional are necessary. However, one
needs at least one term from each of both parts of the functional (the
"causality" part, which has the coefficient λ, and the "fidelity" part). For
instance, the simplest edge linking algorithm is defined by the minimizing
of

$$E_\lambda(K_\lambda) = \lambda \, \text{Card}(\partial K_\lambda) + \text{length}(K_\lambda \setminus K_0) \,.$$

To minimize this functional appears as a continuous version of the Travelling Salesman Problem (Jones [Jon]). The classical "snake method",
found by Kass Witkin and Terzopoulos [KasWT] (and quite close to the
original idea of Montaneri [Mont] and Martelli [Mart]) only keeps the
following terms

$$E_\lambda(K_\lambda) = \lambda \int_{K_\lambda} \left(1 + \text{curv}(\sigma)^2\right) d\sigma - \int_{K_\lambda} |\nabla u(\sigma)|^2 \, d\sigma.$$

We shall list below other instances of the preceding generic functional.

3. Hybrid methods.

They consist in combining both edge and region growing methods. In most cases, either they run first an edge growing and then a region growing, or they run first a region growing and then an edge growing. Of course, the energy associated with these methods is the sum of the energies proposed in the last two sections. Let us consider two typical cases.

R.M. Haralick [Har2] has been one of the first authors to propose a hybrid method. His approach combines edge detection and pure region growing: An edge map is computed by some linear edge operator, labeling thus each pixel as edge or non-edge. Then neighbouring pixels, neither of which are edges, are linked to be in the same region. The results of this technique completely rely on the quality of the edge operator. Haralick detects edges by making at each pixel a cubic spline of the image on a neighbourhood of the pixel. The subsequent region growing may create new edges, but it keeps all edges found in the first stage of the algorithm.

In a recent paper of T. Pavlidis and Y.T. Liow [PavL], one can find the converse method:

> *We present a method that combines region growing and edge detection for image segmentation. We start with a split-and-merge algorithm where the parameters have been set up so that an oversegmented image results. Then region boundaries are eliminated or modified on the basis of criteria that integrate contrast with boundary smoothness, variation of the image gradient along the boundary, and a criterion that penalizes for the presence of artifacts reflecting the data structure used during segmentation (quad-tree in our case).*

Pavlidis and Liow remark that combining two techniques for segmentation produces much better results than what would have been obtained by either technique alone. The general hybrid algorithm is nothing but a slightly more general split-and-merge method. Indeed, the main new tool which is allowed in the process is what can be called "weak splitting". Weak splitting consists of determining by some edge detector an "edge map". This edge map does not necessarily define different regions. Now these edges influence the region growing by favouring the regions whose boundaries contain edges. Thus, the general hybrid algorithm will contain, in arbritary order, all tools of the above analysed methods.

General hybrid split and merge algorithm.

a) *Find an edge map by some edge detector (or define it to be the initial grid).*

    b) *Grow the edges.*

    c) *Grow the regions.*

    d) *Split again the regions.*

    e) *Smooth the edges.*

One can find recent hybrid methods in R. Bajcsy, M. Mintz and E. Liebman [BajLL]; L. Vinet [Vin, part II]; and L. Cohen, L. Vinet, P. T. Sander and A. Gagalowicz [CohVSG], who propose a hybrid linkage region growing method which works simultaneously on both pictures of a stereo pair, therefore making stereo matching, edge detection and region growing collaborate.

4. How to translate a heuristic algorithm into a variational theory.

The reader might ask why such a translation is necessary. The above algorithmic description yields an obvious answer: if one wants to define heuristically each one of the five steps of the preceding algorithm (the "general hybrid split and merge"), one has to find heuristically and independently a lot of parameters (generally thresholds fixing when to merge, when to grow, etc.). As we shall see, the advantage of an *a priori* variational formulation is to reduce the number of parameters and thresholds and to fix them automatically. Indeed, they can be deduced from the weights given to the different terms of the functional to be minimized. We shall illustrate the "translation" method with the most characteristic algorithms mentioned above. The main tool of our translation method is the following proposition, which explains why most split and merge methods are variational.

Proposition 4.1. *Consider a merging method consisting of merging iteratively the regions of a segmentation (O_i). Assume that the criterion for merging O_i and O_j is $a(O_i \cup O_j) - a(O_i) - a(O_j) < 0$ for some function $a(\cdot)$ depending on (O_i). Then the method decreases the energy*

$$E((O_i)) = \sum_i a(O_i).$$

Proof. Let $(O_{i'})$ be the segmentation obtained after the merging of O_i and O_j. Then $E((O_{i'})) - E((O_i)) = a(O_i \cup O_j) - a(O_i) - a(O_j) < 0$. Thus the merging process decreases E.   □

Example 1: Pavlidis' first algorithm (1972).

We start with this algorithm because it seems to have been the first to define the segmentation problem as an optimization problem. The first algorithm of Pavlidis [Pav1] depends on a "merging" threshold λ.

1. Initialization: take all pixels as regions.
2. For every pair of regions O_i and O_j such that $\mathrm{Var}\,(g, O_i \cup O_j) < \lambda$, merge O_i and O_j.

The algorithm is no direct application of our preceding proposition, but corresponds to the <u>constrained optimization</u> problem

$$\min_{\mathrm{Var}\,(g,O_i)<\lambda} \mathrm{Card}\,((O_i)_i),$$

where the minimum is taken on all partitions $(O_i)_i$ of the image domain Ω into connected components O_i. This early formulation had the drawback of giving no control on the smoothness of the boundaries.

Example 2: The Beaulieu and Goldberg clustering algorithm (1989).

We now consider a much more recent example of region growing with a quite close formulation. We shall see that the improvement of the J.M. Beaulieu and M. Goldberg [BeaG] algorithm with respect to the preceding algorithm is to avoid the thresholding effect due to the constraints. Let us quote Beaulieu and Goldberg's description of their algorithm.

A hierarchical segmentation algorithm inspired from hierarchical data clustering and based upon stepwise optimization is now presented. (...)

In a merging scheme, hierarchical clustering starts with N clusters corresponding to each of the N data points, and sequentially reduces the number of clusters by merging. At each iteration, the similarity measures $d(O_i, O_j)$ are calculated for all clusters pairs (O_i, O_j), and the clusters of the pair that minimizes the measure are merged. This merging is repeated sequentially until the required number of clusters is obtained.

The merging criterion is the same as in numerical taxonomy: The regions which are merged are those for which $\mathrm{Var}\,(O_i \cup O_j) - \mathrm{Var}\,(O_i) - \mathrm{Var}\,(O_j)$ is minimal, where $\mathrm{Var}\,(O)$ denotes the variance of the grey level function in the region O (or, more generally, the quadratic distance of the grey level in the region to a polynomial spline of a fixed order).

The subjacent functional E which is minimized by this process is easy to extract. Indeed, Beaulieu and Goldberg give us the amount of energy which has to be "paid" at each merging. They make clear that this amount, $\mathrm{Var}\,(O_i \cup O_j) - \mathrm{Var}\,(O_i) - \mathrm{Var}\,(O_j)$, must be minimal. Define, following the above proposition, the energy of a partition as the sum of the variances of the regions; that is

$$E((O_i)_i) \;=\; \sum_i \mathrm{Var}\,(O_i).$$

Denote by O the partition before merging of O_i and O_j and O' the partition after the merging. Then $E(O') - E(O) = \mathrm{Var}\,(O_i \cup O_j) - \mathrm{Var}\,(O_i) - \mathrm{Var}\,(O_j)$. Thus the Beaulieu and Goldberg algorithm tends to minimize $E(O)$ under the constraint that $\mathrm{Card}\,((O_i)_i)$ is fixed. It solves the problem

$$\min_{\mathrm{Card}\,((O_i)_i)=c} \sum_i \mathrm{Var}\,(O_i)\,,$$

which is the exact converse of the above-mentioned Pavlidis method.

The method of Beaulieu and Goldberg appears to be a path to obtain a partition with small enough cardinality and variance. It is therefore natural to consider it as a method minimizing the energy

$$E((O_i)_i) \;=\; \sum_i \mathrm{Var}\,(O_i) + \lambda\,\mathrm{Card}\,((O_i)_i)\,.$$

Indeed, instead of fixing *a priori* the cardinality of the segmentation, which is a highly arbitrary choice, it seems more natural to weigh the cost of the cardinal by a scale parameter λ. We can reinterpret the Beaulieu and Goldberg algorithm in this context: it is simply a merging method applied to the preceding energy, with a different stopping time; however, the merging method stops when, for any i and j, $\mathrm{Var}\,(O_i \cup O_j) - \mathrm{Var}\,(O_i) - \mathrm{Var}\,(O_j) > \lambda$.

Conversely, does this variational method allow us to find a solution with fixed cardinality? Yes, since it is enough to make the algorithm multiscale. One starts with λ small. Then the cardinality of the initial solution will be high. The scale λ is increased slowly in the merging process until the sought after cardinality is attained.

Notice that the method has the same drawback as the above-mentioned Pavlidis method: no control over the amount or smoothness of the boundaries.

Example 3: The Brice and Fennema "multiregional heuristics" (1970).

The method (see [BriF]) consists of first forming connected components of equal intensity (region growing); refining then with "phagocyte" heuristics; refining with "weakness" heuristics. The phagocyte heuristics consists of evaluating the "weak part" of the common boundary of two regions O_i and O_j, defined as the part across which the jump of the grey level is small. If this weak part is large with respect to the lengths of the perimeters of both regions, then the regions are merged. The weakness heuristics is a variant, consisting of estimating the ratio between the length of the "weak" part of $\partial(O_i, O_j)$ and the length of $\partial(O_i, O_j)$. This method depends on two thresholds, one fixing what a "small" jump of the grey level is, the second fixing which ratio of "weak" boundary is enough to merge two regions.

Let us give a translation of the "weakness heuristic". The merging condition is :

if

$$\beta \,\text{length}\,(\partial(O_i, O_j)) - \text{length}\left\{x \in \partial(O_i, O_j) \middle/ \left|\frac{\partial u_0}{\partial n}\right| < \delta\right\} < 0,$$

then merge O_i and O_j.

Let us again follow Proposition 4.1 and define an energy functional by simply summing the opposites of these merging decision coefficients over all pairs of neighbouring regions O_i and O_j. Denoting by K the union of all boundaries of the O_i, we obtain

$$
\begin{aligned}
E((O_i)_i) \;=\; &-\sum_{i,j} \beta \,\text{length}\,(\partial(O_i, O_j)) \\
&+ \text{length}\left\{x \in \partial(O_i, O_j) \middle/ \left|\frac{\partial u_0}{\partial n}\right| < \delta\right\} \\
=\; &-2\beta \,\text{length}\,(K) + 2\,\text{length}\left\{x \in K \middle/ \left|\frac{\partial u_0}{\partial n}\right| < \delta\right\} \\
=\; &2(1-\beta)\,\text{length}\,(K) \\
&- 2\,\text{length}\left\{x \in K \middle/ \left|\frac{\partial u_0}{\partial n}\right| > \delta\right\}.
\end{aligned}
$$

This achieves the translation of the Brice and Fennema method into a variational method. Indeed, according to the merging condition, two regions will be merged if and only if the energy E decreases by the merging.

Let us now look whether this energy can be rationalized. By rationalization we mean every simplification of the algorithm which avoids thresholding effects. The second term of the merging criterion is essentially defined to measure the contrast of the boundary. This can be done without using any threshold δ by simply setting

$$E(K) = \lambda \, \text{length}(K) - \int_K \left| \frac{\partial g}{\partial n} \right| \, d\sigma.$$

Thus we again find a particular instance of the "general growing" functional which we proposed in the last section. The Brice and Fennema method therefore appears to be a primitive version of the "snakes" method. See page 34.

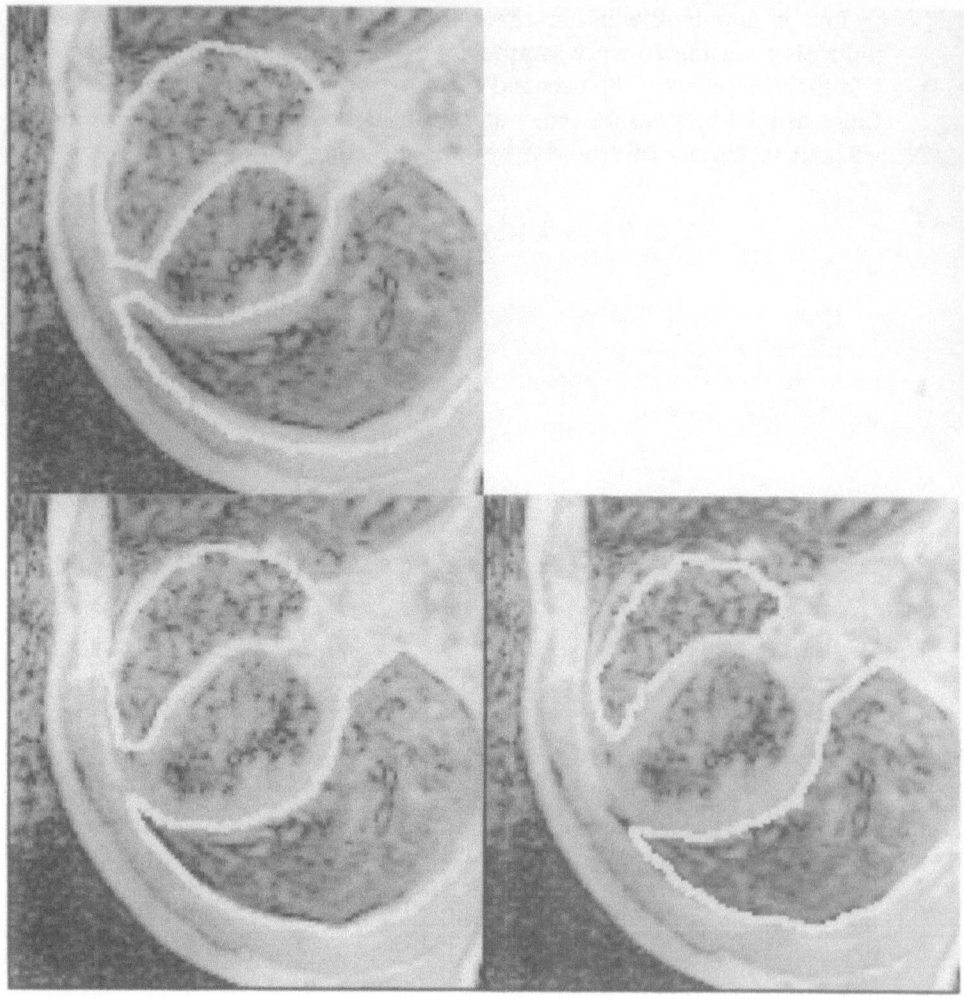

Closed boundaries detection.   Authors: Caselles-Catté-Coll-Dibos

We select an initial contour Γ near to the boundary we want to detect. To this initial contour we associate an image $u_0(x, y)$ such that $\Gamma = \{(x, y) \mid u_0(x, y) = \frac{1}{2}\}$. This image evolves following the equation

$$\frac{\partial u}{\partial t} = g(x)|\nabla u| \left(div\left(\frac{\nabla u}{|\nabla u|}\right) + \nu \right)$$

where g is a function which stops the evolution of the active contour when it reaches the desired contour, and ν is a constant which allows to obtain nonconvex contours. At each iteration the active contour is the level set $\frac{1}{2}$. The pictures presented above are, from upper left to lower right, the image of a heart with the inital contour and the results for 2 and 4 iterations. Computation time: 1 second for each iteration for an image 128×128.

Chapter 4

VARIATIONAL THEORIES OF SEGMENTATION

We have given in the preceding chapter a generic formula for the segmentation problem. This formula was deduced from the algorithmic methods proposed in Computer Vision. We shall now analyse several classical segmentation methods which have been directly defined by a variational formulation. This will give a final confirmation of the well-foundedness of the energy functionals which are discussed in this book.

1. The Nitzberg and Mumford 2.1-D sketch.

The aim of Nitzberg and Mumford [NiM] is to propose a variant of the Mumford and Shah segmentation model allowing regions to overlap. If a region occludes another, this can be interpreted as a "depth" information: one of the objects is in front and the other in back. One good indication for occlusion is given by the so-called "T-junctions". When an object partially occludes another, its edges will cut the edges of the object in back, creating thus a junction of two edges with the shape of a "T". The Gestalt psychology work of Kanisza [Kani] has shown how humans make such reconstructions unconsciously and automatically using T-junctions and other singularities in the image edges. The T-junctions seem to be one of the most relevant informations for ordering objects in space. Indeed, K. Nakayama [NakSS] has explored human perception of data in which various clues for depth, T-junctions, stereo, motion, etc. conflict. His results show in many cases that T-junctions override other clues and are always powerful organizing forces in an image. The information structure obtained by the Nitzberg and Mumford method is mainly two-dimensional (a partition of the images into regions, a segmentation). Now, it also yields an ordering of regions in space; this ordering is not quantitative (a distance information would allow a spatial reconstruction) but just qualitative: for any pair of regions we know which region is in front and which in back. Therefore the name of "2.1" sketch is an intermediate one between the merely 2D segmentation and to the "2.5-D sketch" of David Marr. (The 2.5-D sketch aimed at a faithful quantitative reconstruction of the spatial environment.)

The data structure proposed by Nitzberg and Mumford consists of regions R_i in the domain D of the image corresponding to the distinct objects O_i in front of the lens where the R_i are partially ordered by occlusion between the objects O_i and where the parts of the objects O_i

behind other O_js are reconstructed so that the R_is overlap. In order to compute this data structure, they seek a decomposition of D into overlapping ordered regions $\{R_i\}$ such that

a) The intensity of the image on the visible part of each R_i is as constant as possible.

b) The edges ∂R_i are as short and straight as possible. In particular, invisible edges are reconstructed by the family of curves, called *elastica*, that minimizes a sum of length and the square of curvature.

Figure 4.1 : Two overlapping regions
with four "T-junctions"

The segmentation energy therefore is a variant of the first energy proposed by Mumford and Shah [MumS2] (see Chapter 5, where this energy will be thoroughly studied),

$$E_{2.1}(\{R_i\}, <) =$$

$$= \sum_{i=1}^{n} \left[\mu^2 \int_{R'_i} (g - m_i)^2 dx + \varepsilon \int_{R_i} dx + \int_{\partial R_i \backslash \partial D} \phi(\kappa) d\sigma \right],$$

where

R_i is a family of regions of D such that $\bigcup_i R_i = D$;

" $<$ " is a partial ordering on R_i, that represents occlusion:

$$R_i < R_j \quad \text{if } R_i \text{ is occluded by } R_j \; ;$$

R'_i is the non-occluded or "visible" part of R_i, defined by

$$R'_i = R_i \backslash \{R_j \text{ such that } R_i < R_j\} \; ;$$

m_i is the mean value of g on R'_i;

κ is the curvature on ∂R_i, and ϕ is defined by

$$\phi(\kappa) = \begin{cases} \nu + \alpha\kappa^2 & \text{for} \quad |\kappa| < \dfrac{\beta}{\alpha} \\[2ex] \nu + \beta\kappa & \text{for} \quad |\kappa| \geq \dfrac{\beta}{\alpha} \end{cases}$$

with real parameters α and β.

The first term of the functional $E_{2.1}$ minimizes the variance of the visible parts. The second term corresponds to the assumption that the occluded parts have a minimal area. However, the coefficient ϵ is assumed to be small and this criterion is therefore only used in ambiguous cases. Finally, the third term asks that the boundaries of regions be short and not too curvy. Let us quote the authors to explain the definition of $\phi(\cdot)$:

The purpose of $\phi(\cdot)$ changing from a quadratic to a linear function at $\frac{\beta}{\alpha}$ is to allow us to extend the definition of the second term to a contour which is smooth except at a finite number of infinite-curvature points, or "corners".

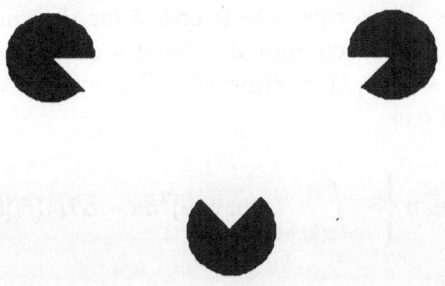

Figure 4.2 : Kanisza's subjective triangle

2. The "structural saliency" of Sha'ashua and Ullman.

Sha'ashua and Ullman's [Sha'U] goal is to construct a *saliency* map which is a representation of an "edge image" emphasizing salient locations. So the initial datum is not a grey level image, but rather a subset of the plane made of possibly ragged and nonsmooth curves. (That is, in fact, a general set with finite Hausdorff length, as we shall see in Chapter 6.) Sha'ashua and Ullman seek to associate a "measure of saliency" denoted by $\Phi(x)$ with each point x in the image. The property which seems

essential in structural saliency is a combination of length and smoothness measured at a particular scale. That is, a measure of saliency would
favour long smooth curves, where the smoothness of a curve is related to
its curvature or its curvature variation. (See Figure 3.1 in the previous
chapter for an illustration of how the perception sees long edges by a an
estimate of curvature variation along broken edges.)

They therefore face the following problems.

1. Defining an appropriate measure $\Phi(x)$ that, when applied to a
   point along a given curve, will increase when the curve increases
   in length and smoothness.

2. A selection problem. The measure $\Phi(x)$ depends on the curve
   passing through x. Since the curves we are considering are either
   continuous or separated by any number of gaps, there will usually be many possible curves to consider. Their approach to this
   problem will be to select the curve that maximizes $\Phi(x)$ over all
   curves passing through x.

Let us describe in a continuous framework how Φ is defined. We assume that an "edge map" $\mathcal{E}$ is given, that is, a set with bounded Hausdorff
measure which can be (in a computational context) assumed to be a finite union of curves. Fix a reference length λ for the curves along which
the "saliency" of x will be computed. Given a curve $c(s)$, parametrized
with its arc length s in $[0, \lambda]$ starting at x (i.e., $c(0) = x$), we define the
saliency measure of c as

$$\mathcal{F}(c) = \int_{c([0,\lambda])\setminus\mathcal{E}} \exp\left\{ -\int_{c([0,\sigma])} \operatorname{curv}(s)^2 ds - \alpha \mathcal{H}^1(c([0,\sigma]) \setminus \mathcal{E}) \right\} d\sigma,$$

where α is a positive attenuation coefficient, $\mathcal{H}^1$ denotes the Hausdorff
length, and curv (s) is the curvature of the curve $c(s)$ at s. The Hausdorff
length will be defined in Chapter 6, but we can understand it here as the
usual Lebesgue measure restricted to the curve.

The saliency measure of x is the upper bound of $\mathcal{F}(c)$, taken over all
smooth curves c starting at x and with length λ. In other terms,

$$\Phi_\lambda(x) = \sup_{\{c \,/\, \mathcal{H}^1(c)=\lambda\}} \mathcal{F}(c).$$

The attenuation coefficient α penalizes "gaps" in the edge map along the
curve c. If, for instance, c is entirely contained in $\mathcal{E}$, then there is no
attenuation of the integrand. It is interesting to notice that, in a discrete

framework, the computation of the structural saliency at each point can be made quick and parallel by using classical algorithms of minimal path search [Mars]. In contrast with many problems in computer vision, there is therefore no combinatorial blow-up in the computation of the saliency map. The Sha'ashua and Ullman theory can be considered as a sound variational formalization of edge growing methods. In the light of what will be developed in Chapters 6 through 12, we can view the Sha'ashua and Ullman method as an attempt to define a local algorithm computing the "most rectifiable part" of a given edge set.

3. The "minimum description length" method of Yvon Leclerc.

Yvon Leclerc [Lec] proposes to interpret the segmentation problem as a partitioning problem, where the criterion for preferring a partition to another should be its "description length". Of course, the measure of the description length must be done according to an *a priori* language. The Minimum Description Length method (MDL) thus defined is related to the classical stochastic approach (MAP: Maximum *a posteriori* probability), but it requires only intuitive information theory concepts. We shall not describe the details of the formalisation but give directly the functional to which Yvon Leclerc is led.

a) Piecewise constant model for a picture:

$$E(u, K) = \frac{1}{\sigma^2} \sum_{i \in I} (u(i) - g(i))^2 + \beta \sum_{i \in I} \sum_{j \in N(i)} (1 - \delta(u_i - u_j)),$$

where g is the picture defined on the set of pixels I, σ is the variance of the white noise added to the picture, δ is the Kronecker symbol, $N(i)$ is the set of the neighbouring pixels of i, β is inversely proportional to the average length of the boundary of a region.

b) Piecewise polynomial model for a picture: In this case, the first term remains the same, but it is augmented with the number $n(i)$ of nonzero Taylor coefficients at i, multiplied by the number α of bits used for coding such a coefficient. The "length" term is replaced by a term $m(i, j)$ proportional to the number of jumps of the Taylor coefficients of u between i and its neighbours. Thus the functional becomes

$$E(u, K) = \frac{1}{\sigma^2} \sum_{i \in I} (u(i) - g(i))^2 + \alpha \sum_{i \in I} n(i) + \beta \sum_{i \in I} \sum_{j \in N(i)} m(i, j).$$

In Leclerc's theory, the coefficient α is a scale parameter since it is inversely proportional to the average surface of a region, and β is another scale parameter inversely proportional to the average length of a region boundary. We shall prove in Chapter 5 that both scale parameters can be reduced to one, because in an optimal segmentation for the Mumford and Shah piecewise constant model, the length and the area of a region are bounded by isoperimetric and inverse isoperimetric inequalities. The Leclerc model looks like a variant of the Mumford and Shah general model, where piecewise analytic functions are replaced by piecewise polynomial functions. However, the Mumford and Shah model allows "open" boundaries to arise, that is, boundaries which do not separate two different regions. The piecewise constant and the piecewise polynomial model instead are partitioning models.

4. Algorithms for minimizing the general Mumford and Shah energy.

Blake and Zisserman have devoted a book [BlakZ] to the problem of minimizing the Mumford and Shah energy, which they call the "weak membrane model". We shall limit ourselves to give an idea of the "Graduated Non Convexity" Algorithm. This algorithm is the most successful for minimizing the general Mumford and Shah energy. Its principle, a "coarse to fine homotopy method" has since been followed implicitly or explicitly by most authors. There is no loss of generality in describing the algorithm in the case of a one-dimensional datum instead of a picture. The principles of the algorithm are quite the same. Let us consider the Mumford and Shah energy for analysing a one-dimensional datum $d(x)$ defined on a real interval I:

$$E(u, K) = \int_I [(u(x) - d(x))^2 + u'(x)^2]\, dx + \mathrm{Card}\,(K).$$

(For the sake of simplicity we omit the scale parameters.) K denotes the zero-dimensional edges, a set of points which can be interpreted as the true "jumps" of d. The function u is a piecewise smooth approximating function for d, with jumps at the points of K. In order to discretize this energy, one defines a grid of points on I: $i = 1, 2, \ldots, n$.

Define

$$g^0(t) = \begin{cases} t^2 & \text{if } t \le 1 \\ 1 & \text{if } t \ge 1. \end{cases}$$

Then a discrete version of E is

$$E(u) = \sum_{i=1}^{n} (u_i - d_i)^2 + g^0(u_i - u_{i-1}),$$

and K is defined as the set of the i for which $|u_i - u_{i-1}| > 1$. This energy is nonconvex and can have a lot of local minimizers. The idea of Blake and

Zisserman is to convexify E in order to get a global minimizer. Then this minimizer is used as a starting point for a less convex version of E. Thus a series of more and more nonconvex versions of E which are successively minimized is built. Of course, nothing ensures that this method will provide a global minimum of E. It simply is a sound heuristics. Define

$$g^p(t) = \begin{cases} t^2, & \text{if } |t| < \dfrac{1}{r} \\ 1 - \dfrac{(|t|-r)^2}{4p}, & \text{if } \dfrac{1}{r} \leq |t| < r \\ 1, & \text{if } r \leq |t| \end{cases}$$

where $r = \sqrt{4p+1}$.

Set the smoothed versions of E to be

$$E^p(u) = \sum_{i=1}^{n}(u_i - d_i)^2 + g^p(u_i - u_{i-1}),$$

Then it is shown that if $p \geq 1$, E^p is convex. Thus a global minimizer can be easily found for E^p if $p = 1$. Note that $E = E^0$. So the progressive nonconvexity method consists in letting p decrease from 1 to 0.

N.K. Nordström [Nor] also studied the problem of defining procedures for minimizing the Mumford and Shah functional, with a term added, which is the number of curves of the segmentation. This addition, also studied in Richardson's thesis [Ric], makes very easy the existence results for the minimization problem, as we shall see in Chapter 15.

The proposed algorithm is general, however, and admits *a priori* any kind of configuration for edges. Its steps are:

a) Initialize the segmentation using some edge detector.

b) Move the obtained edges so that the energy decreases.

The Nordström algorithm is a "hybrid method", close to the "snakes" method in the edge moving part, and therefore different from the Blake and Zisserman approach. Tom Richardson [Ric] also studied several properties of the Mumford and Shah functional and presented experiments about its minimization in his thesis. He used finite elements to discretize the Mumford and Shah functional with the Γ-convergence approach of Ambrosio and Tortorelli [AmbT]. This last method can also be interpreted as "coarse to fine homotopy method": like the GNC algorithm of Blake and Zisserman, it consists of defining a rough version of the functional (with "thick" edges) and to use the minimum of this version as an initial point to minimize a finer version of the functional, and so on. Ambrosio and Tortorelli proved the convergence of a special class of "coarse functionals" approximating the Mumford and Shah energy. Chambolle [Chamb] has obtained the same convergence result for a wide class of discretizations used in computer vision and including the stochastic methods proposed by Geman and Geman, the Blake and Zisserman above-mentioned algorithm, etc. Figure 4.3 shows a Mumford and Shah segmentation obtained by Chambolle with an algorithm which is a fast variant of the Graduated Nonconvexity algorithm.

5. Affine invariant segmentation.

We shall end this chapter with an interesting alternative to the Mumford and Shah functional which makes it invariant with respect to affine anamorphoses of the image. The classical Mumford and Shah algorithm is euclidian invariant, which means that whenever an image is rotated or translated, then so are its minimal segmentations. Now, the analysis of natural scenes leads to projective transforms of plane objects in view in the scene. Not much has been done to our knowledge with projective invariant segmentation. A simplified case of projective invariance is the *affine invariance*, which corresponds to the case of plane objects seen at some distance. If an image $g(x)$ is presented to the eye or to a camera at some distance and slanted, then the resulting image is $g(Ax)$ where A is some affine transform. We would like to obtain $u(Ax), AK$ as an optimal segmentation pair for $g(Ax)$, whenever $u(x), K$ is optimal for $g(x)$ and $det(A) = 1$. Ballester, Caselles and González [BCG] proposed to do that by minimizing the functional

$$E(u, K) = \int (u - g)^2 dx + \int \int_{KxK} |d\nu(x) \wedge d\nu(y)|,$$

where K is made of rectifiable curves, $\nu(x)$ denotes the unit normal to K at x, and $u(x)$ is piecewise constant on the complementary of K. It is easily seen that $E(u(Ax), AK) = E(u(x), K)$ whenever $det(A) = 1$. The authors prove existence of a minimizing segmentation made of a finite number of curves with a method close the method that we develop in Chapter 5 for the piecewise constant Mumford and Shah model. One of the remarkable properties of the affine invariant functional is the allowance of long and thin regions, because for instance a thin rectangle is affine equivalent to a square with same area. This property makes affine segmentation useful for satellite images, where roads and long structures in general are meaningful. (See Figure 4.3, where a region growing similar to the algorithm described in Section 5.4 of Chapter 5 has been applied, with the affine invariant energy functional as merging criterion.)

Top right, the original image (GRECO). Bottom right, the
reconstructed datum u. Left, a superposition of the reconstructed datum u and the edge set K (in white).

This figure shows a low-energy Mumford and Shah segmentation. Various methods exist
to reach such low-energy states. The method used here by Antonin Chambolle [Cham1,2] is a
hybrid method: first the two-dimensional part of the energy is minimised (this results in the
minimisation of a convex functional). Then the main edges are detected by a standard method
(extrema of gradient). This set of edges K_1 being fixed, the two-dimensional energy is once
again minimised on $\Omega \setminus K_1$, leading to the detection of a set of finer edges, K_2, etc. On this
image, there are 50 such iterations.

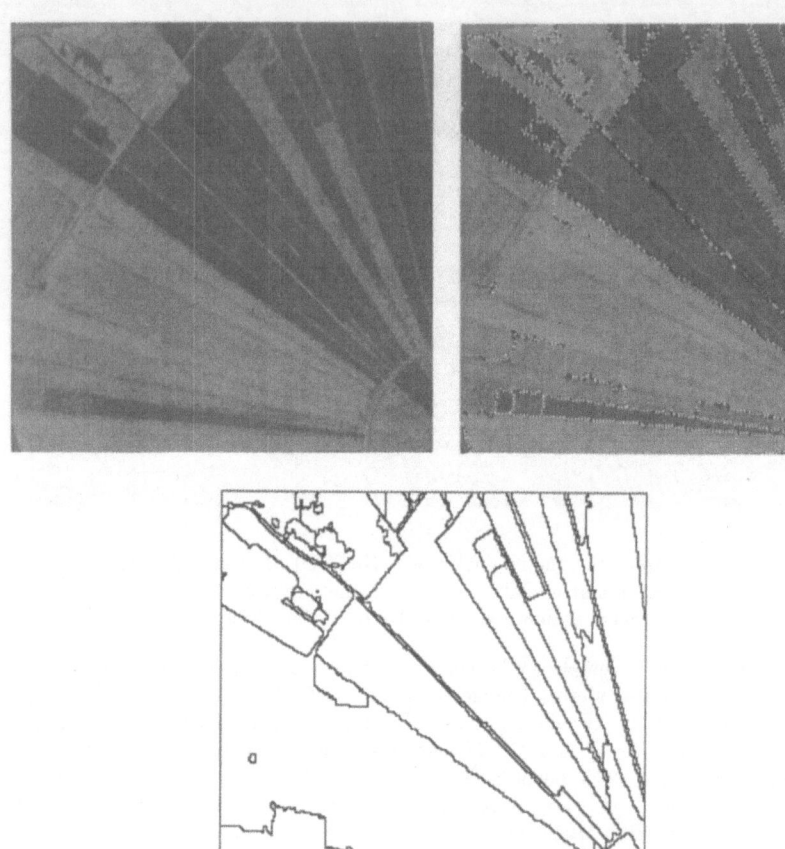

<u>Authors:</u> C. Ballester, V. Caselles, M. González [BCG].

Affine invariant segmentation of a SPOT image, obtained by region growing with the affine invariant energy

$$E(u, K) = \int (u - g)^2 dx + \int \int_{K \times K} |d\nu(x) \wedge d\nu(y)|$$

($\nu(x)$ is the normal at x to K.)

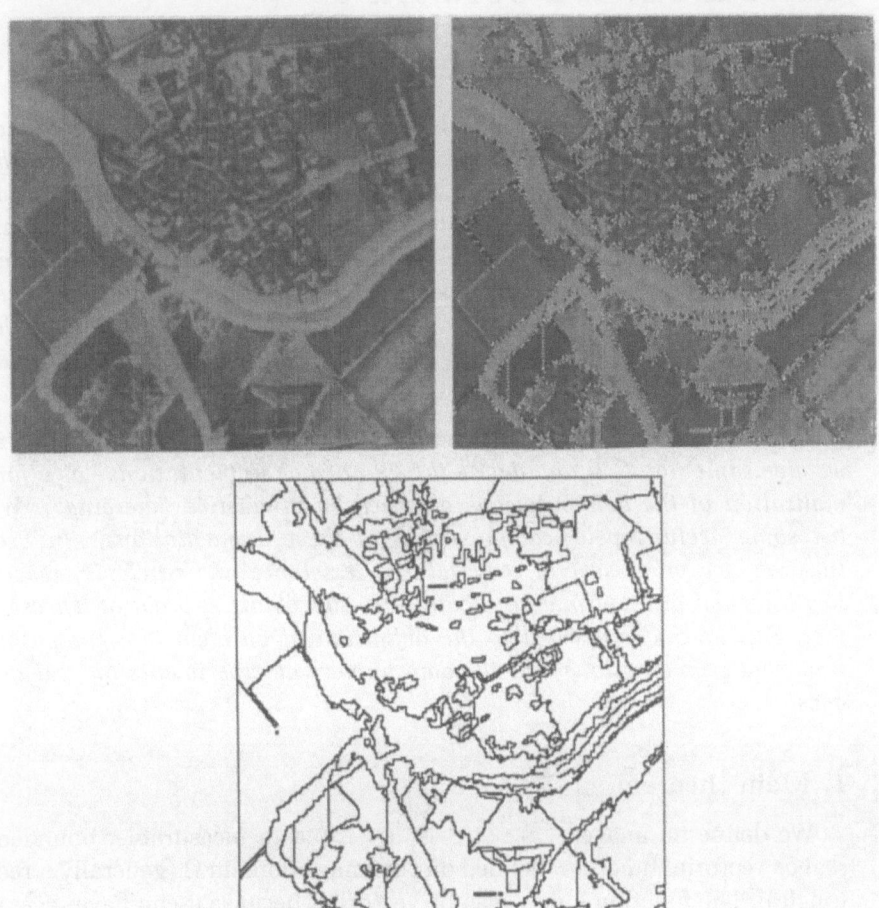

<u>Authors:</u> C. Ballester, V. Caselles, M. González [BCG].

Affine invariant segmentation of a SPOT image, obtained by region growing with the affine invariant energy

$$E(u,K) = \int (u - g)^2 dx + \int \int_{K \times K} |d\nu(x) \wedge d\nu(y)|$$

($\nu(x)$ is the normal at x to K.)

Chapter 5

THE PIECEWISE CONSTANT MUMFORD AND SHAH MODEL: MATHEMATICAL ANALYSIS

In this chapter, we start with the mathematical proofs and show that the minimizing of the "simplest" segmentation energy discussed in the first chapters, the Mumford and Shah piecewise constant model, yields properties sought by most segmentation devices. Indeed, the most primitive segmentation tool, the "merging", applied to this variational model, is enough to ensure a small (compact) set of possible segmentations, with no small regions and no thin regions. Uniform a priori estimates for the size and number of the regions can be given for all segmentations obtained by exhaustive "merging".

In Section 5.1, we state the main existence theorem. In Section 5.2, we give some notation and define the "2-normal segmentations" as a formalization of the segmentations obtained by exhaustive "merging". We list some useful topological properties of these segmentations. In Section 5.3, we prove several compactness, existence and regularity results for 2-normal and optimal segmentations and obtain a proof of Theorem 5.1. Section 5.4 is devoted to the definition of an algorithm computing 2-normal segmentations and to some numerical experiments on real images.

1. Main theorem.

We define an image $g(x) : \Omega \subset \mathbb{R}^2 \mapsto \mathbb{R}^N$ as a measurable, bounded, real or vectorial function, defined on the image domain Ω (generally a rectangle). The function g may also be vectorial, because it can have several "channels" $g_1, ..., g_N$. If there only is one channel ($N = 1$), this channel generally corresponds to the grey level or brightness of the image. In case of several channels, each one denotes the density of some attribute, which can be for instance colour (red, green, blue) or even "texture density", as proposed by authors concerned by texture segmentation. The fact that g is vectorial does not change anything in the theorems and proofs that we state in this chapter. The only underlying hypothesis for making the Mumford and Shah functional sound is that these channels have been defined in order to characterize the similarity of points of the picture and therefore the autosimilarity of regions. We seek a strong segmentation, that is, a *partition* of the rectangle Ω into a finite set of regions, each of which corresponds to a part of the image where g is as constant as possible. Moreover, we wish to compute the region boundaries and to control their regularity.

As we have seen through the preceding chapters, the simplest energy functional associated with such a generic description is the Mumford and Shah variational model [MumS2]. According to this model the segmentation (u, K) should be obtained by minimizing the functional

$$(1) \qquad E(u, K) = \int_{\Omega \setminus K} (u - g)^2 dx + \lambda l(K) \,,$$

where K is a union of boundaries in Ω with length $l(K)$, and u is piecewise constant on $\Omega \setminus K$. The constant λ represents the scale parameter of the functional and measures the amount of boundary: If λ is low, a lot of boundaries are allowed and we shall get a fine segmentation. As λ increases, the segmentation gets coarser.

This functional represents the simplest compromise between accuracy of the regions and parcimony of the boundaries. It is also the only one for which a complete mathematical analysis and some computational theory are available. Our purpose is to prove the following theorem, due to Mumford and Shah [MumS2]. We denote by $osc(g)$ the oscillation of g, defined by $osc(g) = sup(g) - inf(g)$.

Theorem 5.1. *Let g be a measurable bounded function in Ω. Then the minimum of $E(u, K) = \int_{\Omega} (u - g)^2 dx + \lambda l(K)$ is attained at some K. Moreover, the minimal boundary sets have the following geometric property: either the points of K are regular, C^1 and with curvature bounded by $\frac{8osc(g)^2}{\lambda}$, or the singular points are of two types, namely, triple points where three branches meet with 120° angles and boundary points where K meets the boundary of Ω at a 90° angle.*

Of course, the preceding theorem will make sense only when the terms "segmentation" and "length" have been well defined, which is the aim of the next section.

2. Definition and elementary properties of segmentations.

In this subsection, we define segmentations as finite unions of rectifiable Jordan curves and we give bounds on the number of "edges" and "crossings" of such segmentations in terms of the number of "regions".

We begin with a list of definitions and properties of rectifiable curves and segmentations made of a finite number of rectifiable curves. More will

be proved in Chapters 6 to 12 on the subject of rectifiability. This chapter
will be self-contained, however, and proceed with elementary definitions
and proofs.

- A *rectifiable curve* is a map $c(t)$ from $[0,1]$ into $\mathbb{R}^2$ such that
  $l(c) =: sup(|c(a_2)-c(a_1)|+...+|c(a_n)-c(a_{n-1})|)$ is finite, where the
  preceding supremum is taken among all finite increasing sequences
  $a_1, ..., a_N$ of real numbers in $[0,1]$. The real number $l(c)$ is called
  length of c. Let c be a rectifiable curve and let $\sigma(t) = \frac{l(c_t)}{l(c)}$, where
  c_t is the restriction of the curve c to the interval $[0,t]$. Then we
  can reparametrize c by setting, for any s in the interval $[0,1]$,
  $c^1(s) = c(\sigma^{-1}(s))$. One deduces easily from the definition of l
  and the triangular inequality that $|c^1(s) - c^1(s')| \leq l(c)|(s - s')|$.
  Conversely, if $c(t)$ is a curve with Lipschitz constant L on the
  interval $[0, 1]$, then c is rectifiable and its length is less than
  L. (The proof is obvious with our definition of the length.) In
  the following, we therefore always assume that the considered
  rectifiable curves have been parametrized with length. We shall
  always identify a rectifiable curve c and its range so that "c"
  means both the curve and its range in the plane.

- *Ascoli-Arzela Theorem*: Let c^k be a sequence of functions which
  are uniformly Lipschitz on the interval $[0,1]$ and such that the set
  $c^k(0)$ is bounded. Then one can extract a subsequence of c^k which
  converges uniformly to a function c with at most the same Lipschitz constant. Applied to a sequence of curves c^k with $l(c^k) \leq L$
  and $c^k(0)$ bounded, this theorem asserts that a subsequence of the
  c^k converges uniformly to some rectifiable curve c with length less
  or equal to L.

- *Curvature*: Let $c(t)$ be a C^1 rectifiable curve parametrized with
  length. Then one checks easily that $|c'(t)| = 1$. If $c(t)$ is twice
  differentiable at t, we define the curvature at t, $curv(t)$, as the
  real positive number $|c''(t)|$.

- A *Jordan curve* is a continuous curve, such that for all $\sigma, \sigma' \in$
  $[0, 1]$: $c(\sigma) \neq c(\sigma')$ if $\sigma \neq \sigma'$ unless $\{\sigma, \sigma'\} = \{O, 1\}$; if $c(0) =$
  $c(1)$ then the Jordan curve is said to be *closed*. If $c(0)$ and $c(1)$
  are different, we call them *tips* of the Jordan curve. All other
  points are called *interior points* of the Jordan curve.

- A *segmentation* is a union of a finite set of rectifiable curves.

- *Length of a segmentation K*: We define $l(K)$ as the infimum of
  the lengths of all countable sets of rectifiable curves whose union

is K. If, for instance, K is the union of a set of rectifiable curves meeting only at an countable set of points, it is easily seen that $l(K)$ is exactly the sum of the lengths of the curves.

- *Regions of a segmentations*: the connected components of $\Omega \setminus K$. We shall denote them by $(O_i)_i$. The *two-dimensional Lebesgue measure* of O_i is denoted by $|O_i|$.

- *Convergence of a segmentation.* We shall say that a sequence of segmentations K_n tends to a segmentation K if each K_n is a union of Jordan curves (c_i^n) for $1 \leq i \leq k$, if each (c_i^n) tends uniformly to some curve c_i, and if K is the union of the ranges of the c_i.

- The *energy $E(K)$*: If we fix the boundaries K, then the corresponding minimal u is completely defined by the fact that its value on each connected component O of $\Omega \setminus K$ is equal to the mean value of g on this connected component. Thus we shall always assume in the following that with each K is associated this unique value $u = \frac{1}{|O|} \int_O g(x) dx$. Therefore we shall write $E(K)$ instead of $E(u, K)$.

- *Common boundary* of two regions O_i and O_j: we denote it by $\partial(O_i, O_j)$. It is contained in K. If $i = j$, ∂O_i denotes the boundary of O_i.

- *Isoperimetric inequality* in $\mathbb{R}^2$ and Ω: denote by O a region in $\mathbb{R}^2$ whose boundary is an countable union of rectifiable curves. Then

$$l(\partial O) \geq 2\sqrt{\pi} \cdot \sqrt{|O|}.$$

In the case of Ω, the same kind of inequality holds for the *relative boundary* of O in Ω, $\partial O \cap \Omega$, with a smaller constant C:

$$l(\partial O \cap \Omega) \geq C \cdot \sqrt{|O|},$$

if we in addition assume that $|O| \leq \frac{1}{2}|\Omega|$. If, for instance, Ω is a rectangle, then $C = \sqrt{2\pi}$. Since there is no ambiguity, Ω being fixed, we still denote by ∂O the relative boundary of O in Ω.

- *1-normal segmentations*: A segmentation will be called *1-normal* if it is made of a finite number of rectifiable Jordan curves, meeting each other and $\partial\Omega$ only at their tips and if *each Jordan curve separates two different regions*. This last property ensures that the number of regions of the segmentation is finite. Notice that if a Jordan curve does not separate two different regions, then we can decrease the energy by simply removing the curve. Thus

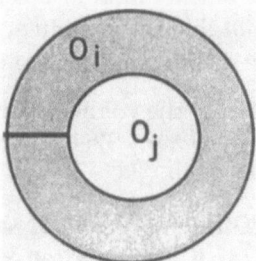

<u>Figure 5.1</u> : a segmentation which is not 1-normal

this property is no restriction. We finally impose that each tip is
a common tip of at least three Jordan curves. (If only two Jordan curves meet at a tip, we can "merge" them into a single one
without changing the segmentation.)

- *Crossing points of a 1-normal segmentation*: all the points of K
  where at least three Jordan curves have a common tip, or where
  a Jordan curve meets $\partial\Omega$ at a tip.

- *Edges of 1-normal segmentation*: Each one of the Jordan curves
  defining a 1-normal segmentation. The edges can be equivalently
  defined as all the connected components of the common boundaries $\partial(O_i, O_j)$.

- *2-normal segmentations*: A segmentation K will be called 2-
  normal if for every pair of regions O_i and O_j, the new segmentation K' obtained by merging these regions satisfies $E(K') >
  E(K)$. (Compare to the definition of 1-normality.)

The following result is proved for instance in [Ale].

Lemma 5.2 (Jordan curve Lemma). *Every closed Jordan curve c
divides the plane into exactly two connected components, one bounded,
"enclosed by c", and the other one, unbounded.*

More generally, we shall say that a set A is *enclosed* by a set B if A is
contained in a bounded connected component of the complementary of
the set B. The proof of the next lemma is a little bit informal and uses
some topological properties of $\mathbb{R}^2$ which are intuitively obvious. However,
they all can be formalized by an accurate use of the Jordan curve Lemma.

Lemma 5.3. *Let K be a 1-normal segmentation with α regions. Then K can be decomposed into a union of $\alpha - 1$ Jordan curves meeting only at a finite set of points.*

Proof. We shall show that we can eliminate one Jordan curve from the set K in such a way to reduce to $\alpha - 1$ the number of the regions and to still obtain the segmentation 1-normal. Then the statement will clearly follow by induction. Take two regions O_i and O_j such that $l(\partial(O_i, O_j)) > 0$. If $\partial(O_i, O_j)$ is connected, it is a Jordan curve and we merge both regions. Then no piece of the common boundary of both regions is left and the segmentation remains 1-normal. Otherwise, the union of O_i and O_j encloses at least one region O_k such that $l(\partial(O_i, O_k)) > 0$. Then we iterate the process with O_i and O_k instead of O_i and O_j. Since the number of regions is finite, this process must stop after a finite number of steps. More precisely, after less than $\alpha - 2$ steps, there are two regions left and the segmentation is 1-normal. Then there is a single Jordan curve left. □

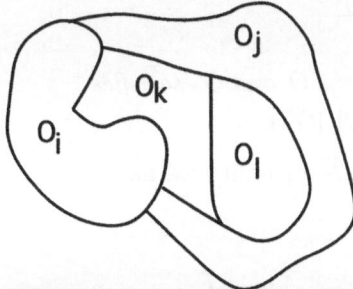

<u>Figure 5.2</u> :   Two regions enclosing a pair of regions

Lemma 5.4. *Let α be the number of regions of a 1-normal segmentation K, β the number of edges and γ the number of crossing points. Then*

$$\gamma \leq 2 \cdot (\alpha - 1) \quad and \quad \beta \leq 3 \cdot (\alpha - 1) - 2 \ .$$

Proof. Consider the proof of the preceding lemma. The first curve to be removed, $\partial(O_i, O_j)$, contains at most two crossing points of K: its tips if it is an open Jordan curve. So the merging of O_i and O_j eliminates at most two crossing points. By iterating the process the first inequality follows. Let us now analyse how the number of edges decreases by the merging. If $\partial(O_i, O_j)$ is a closed Jordan curve, it is the only edge to disappear. If $\partial(O_i, O_j)$ is an open Jordan curve then there are two crossing points at its tips. If one of these crossing points meets more than three edges of K, the other edges are not modified. But if there are only three edges at this crossing point, it will vanish when $\partial(O_i, O_j)$ is removed, and the other two edge curves will merge into one. Thus the merging decreases β of at most 3, and iterating the process yields the second inequality. $\square$

3. Properties of segmentations obtained by merging. Proof of the main theorem.

In this section, we develop a constructive strategy for proving Theorem 1, as well as meaningful lemmas concerning the set of 2-normal segmentations. We consider an arbitrary sequence K_n of 2-normal segmentations and give a priori lower bounds on the areas of the connected components of $\Omega \setminus K_n$. We deduce upper bounds on the number of Jordan curves defining K_n and on the number of crossing points of K_n. We conclude then the proof of Theorem 1 by showing that all these estimates pass to the limit along a minimizing sequence made of 2-normal segmentations.

Lemma 5.5. *Let α be the number of regions of a 2-normal segmentation (u,K). Then $\alpha \leq \frac{288 \, |\Omega| \, \text{osc} \, (g)^4}{C^2 \lambda^2}$.*

Step 1. *Every pair of regions O and O' satisfies*
$$l(\partial(O,O')) \leq \frac{\text{osc} \, (g)^2}{\lambda} \cdot \min(|O|, |O'|).$$

Let $K' = K \setminus \partial(O,O')$. By 2-normality we have

$$
\begin{aligned}
0 &\leq E(K') - E(K) \\
&= \int_{O \cup O'} (u''-g)^2 - \int_O (u-g)^2 - \int_{O'} (u'-g)^2 - \lambda l(\partial(O,O')) \,,
\end{aligned}
$$

where u, u' and u'' denote the mean values of g on O, O' and $O \cup O'$. We get, supposing for example that $|O| \leq |O'|$,

$$\lambda l(\partial(O,O')) \leq \int_O \left[(u'-g)^2 - (u-g)^2 \right] \leq \min(|O|,|O'|) osc(g)^2.$$

Step 2. *For every region O, denote by $N(O)$ the number of neighbouring regions. Then $N(O) \geq \dfrac{C\lambda}{\sqrt{|O|} \, \text{osc} \, (g)^2}$, where C is the isoperimetric constant in Ω.*

Call O_j a neighbouring region of O. The fact that O and O_j cannot be merged without increasing the energy E implies by the preceding step that $\lambda l \left(\partial \left(O, O_j \right) \right) \leq |O| \; \text{osc} \, (g)^2$. Thus by adding these inequalities for all neighbours of O, we obtain $\lambda l(\partial O) \leq N(O)|O|$ osc $(g)^2$. We conclude by applying to O the isoperimetric inequality in Ω.

Step 3. The union of all regions O_i is equal to the rectangle Ω and therefore $\sum_i |O_i| = |\Omega|$. Thus the number of O_is satisfying $|O_i| \leq \frac{2}{\alpha}|\Omega|$ is larger than $\frac{\alpha}{2}$. Let us now apply the result of Step 2 to all of these O_is. Each one of them has at least $\frac{C\lambda}{\sqrt{|O_i|}\,\mathrm{osc}\,(g)^2} \geq C\lambda \cdot \sqrt{\frac{\alpha}{2\,|\Omega|}} \cdot \frac{1}{osc(g)^2}$ neighbouring regions. Consequently the number β of pairs of adjacent regions, and therefore of edges, satisfies

$$\beta \geq \frac{\alpha}{4} \cdot C\lambda \sqrt{\frac{\alpha}{2\,|\Omega|}} \cdot \frac{1}{\mathrm{osc}\,(g)^2} = C\lambda 2^{-\frac{5}{2}} \cdot \frac{\alpha^{\frac{3}{2}}}{\sqrt{|\Omega|}} \cdot \frac{1}{\mathrm{osc}\,(g)^2} \cdot$$

By Lemma 5.4, $\beta \leq 3\alpha$. Thus $C\lambda 2^{-\frac{5}{2}} \frac{\alpha^{\frac{3}{2}}}{\sqrt{|\Omega|}\,\mathrm{osc}\,(g)^2} \leq 3\alpha$ and we obtain the announced relation. □

Remark: Elimination of small regions.

It is easy to deduce from the preceding proof a lower bound on the area of each region of the segmentation. Indeed, take any region O of the segmentation. By Step 2 of Lemma 5.5, the number of neighbouring regions O_j is at least $N(O) \geq \frac{C\lambda}{\sqrt{|O|}\,\mathrm{osc}\,(g)^2}$. Thus the number α of regions of the segmentation is at least $\frac{C\lambda}{\sqrt{|O|}\,\mathrm{osc}\,(g)^2}$. By using the upper bound for α given by Lemma 5.5, we get $\frac{288\,|\Omega|\,\mathrm{osc}\,(g)^4}{C^2\lambda^2} \geq \frac{C\lambda}{\sqrt{|O|}\,\mathrm{osc}\,(g)^2}$. Thus the area of O is bounded from below by a positive constant only depending on g, λ and Ω. Therefore a merging method based on the minimizing of the energy $E(K)$ will spontaneously eliminate the small regions. It is also easy to deduce from the above estimates that the regions are not too "thin", that is, satisfy an "inverse isoperimetric inequality", $|O|^{\frac{1}{2}} \geq C \cdot l(\partial O)$ for a constant C depending only on g, λ and Ω.

The next result gives a mathematical explanation of the efficiency of region growing methods. In order to make the proof easy, we need to define an intrinsic distance between sets. If K is a subset of Ω, we denote by K^ε the set of all x in Ω such that $d(x, K) \leq \varepsilon$. The *Hausdorff distance* between two subsets K and L of Ω is defined as the infimum of all positive numbers ε such that $K \subset L^\varepsilon$ and $L \subset K^\varepsilon$.

Lemma 5.6. *For every sequence K_n of 2-normal segmentations, there exists a subsequence converging for the Hausdorff distance to a segmentation K such that*

$$E(K) \leq \liminf_n E(K_n) \,.$$

Proof. Notice that the limit segmentation K is not necessarily 2- (or even 1-) normal. The proof of the announced compactness property is essentially based on the fact that the number of edges of any 2-normal segmentation is now bounded from above by the preceding estimates (Lemmas 5.4 and 5.5). By the Ascoli-Arzela theorem, each of the edges can be supposed to converge to a limit rectifiable curve if we extract an ad hoc subsequence. The limit segmentation is then defined by this finite set of rectifiable curves. When passing to the limit, we shall prove that neither the integral part of the energy nor the length of the edges can increase, and therefore this limit segmentation has an energy smaller than the inf limit of the energies of the sequence.

Note that by our conventions, u_n and u are automatically defined from K_n and K. Fix $\varepsilon \geq 0$. For n large enough the Hausdorff distance between K and K_n satisfies $d(K_n, K) \leq \varepsilon$. Then K_n is contained in K^ε, and every connected component O_i^ε of $\Omega \setminus K^\varepsilon$ is included in one component O_i^n of $\Omega \setminus K_n$. Thus u_n is locally constant on every connected component of $\Omega \setminus K^\varepsilon$. Since these connected components are countable, we can extract a subsequence of u_n, still called u_n, which converges pointwise in $\Omega \setminus K^\varepsilon$. By a diagonal selection argument, one can make ε tend to zero and extract another subsequence u_n converging pointwise everywhere in $\Omega \setminus K$. Let $u = \lim u_n$. Clearly u is constant in every connected component of $\Omega \setminus K$. Indeed, taken two points x, y of such a component, one has x and y in the same component of $\Omega \setminus K^\varepsilon$ for ε small enough and therefore $u_n(x) = u_n(y)$ for large n, that is finally $u(x) = u(y)$. Thus (u, K) is a segmentation of g, and moreover by Fatou's lemma one has $\int_\Omega (u-g)^2 \leq \liminf_n \int_\Omega (u_n-g)^2$. By using the Ascoli-Arzela theorem we also obtain that the lengths of the curves of the segmentation can only decrease when passing to the limit. $\quad\square$

Lemma 5.7. *There exists a 2-normal segmentation realizing $\min_K E(K)$, this minimum being taken among all 2-normal segmentations.*

Proof. From the previous lemma, it is clear that we can choose a sequence K_n of 2-normal segmentations with the following properties:

a) $K_n = (c_n^k)_k$ is made of a constant number of Jordan curves $(c_n^k)_k$ from $]0,1[$ into Ω. Their tips are a finite set of points, A = $(a_n^i)_i$, the crossing points of K_n, which may belong to the boundary of Ω.

b) Each sequence $(a_n^i)_n$ converges to a point a^i of the closure of Ω.

c) Each sequence of curves $(c_n^k)_n$ converges uniformly to a rectifiable curve c^k contained in the closure of Ω.

d) The sequence $E(K_n)$ tends to $\inf_K E(K)$, this infimum being taken among all 2-normal segmentations.

Let K be the segmentation defined as the union of the ranges of the $(c^k)_k$. We shall prove that K is 2-normal. This is equivalent to proving that the only crossing points of K are the a_is. Assume by contradiction that a new crossing point a appears as we pass to the limit. We first look at the case where a is in Ω. Consider a disk D with center a and radius ε. Fix ε very small so that ε^2 is much smaller than ε. That means that ε is chosen small enough to allow us to neglect the terms less than ε^2 in all the estimates which follow. Those terms will appear as areas of regions contained in a disk of radius ε and will be omitted when compared to length terms of order $C\varepsilon$, where C denotes various absolute constants. Assume moreover that $\varepsilon \leq \min_i |a - a_i|$, and take n large enough to ensure

$$E(K_n) \leq \liminf_n E(K_n) + \varepsilon^2 \quad , \quad |a_i^n - a_i| \leq \varepsilon^2 \quad , \quad |c_n^k - c^k| \leq \varepsilon^2 \ .$$

Consider all maximal pieces of curves of K contained in D. Let us now replace each of these pieces of curves by an affine curve. Since the c_n^k do not cross in D, these affine curves do not cross. (This fact is an easy consequence of the Jordan Curve Lemma.) By this process the connected components of $\Omega \setminus K$ remain unchanged outside D and the length of the curves goes down. Call K' the new obtained segmentation. We still have $E(K') \leq \min_K E(K) + C\varepsilon^2$. (In fact we can take $C = (4\pi + 1)osc(g)^2$.) Now consider two affine curves $[u, v]$ and $[x, y]$ of K' which have been substituted to two parts of curves of K_n passing at a distance less than ε^2 from a. Since a is a crossing point of K, such curves exist for large n. The lengths of the corresponding pieces of curves are greater than $2\varepsilon - 2\varepsilon^2$ since they pass at a distance less than ε^2 from the center a of D. Moreover, when replaced by affine curves, their lengths cannot decrease more than $C\varepsilon^2$. Thus $[u, v]$ and $[x, y]$ are in fact nearby diameters of D which do not cross. By exchanging u and v if needed, we easily obtain that $|u - x|$ and $|v - y|$ are less than $C\varepsilon^2$ for some other absolute constant C. The contradiction comes by modifying K' as follows: We add to K' the segments $[u, x]$ and $[v, y]$ and remove *one* of the segments $[u, v]$ or $[x, y]$. This does not modify the connected components outside D. Thus we have subtracted from $E(K')$ a length of order 2ε, which is a contradiction. If now a new crossing point appears in K on the boundary of Ω, the proof is essentially the same and is left to the reader. $\square$

Lemma 5.8. *Let K be an optimal 2-normal segmentation. Then the Jordan curves of K are twice differentiable and their curvature is bounded by* $8\frac{osc(g)^2}{\lambda}$.

Proof. Let $c(s)$ be a parametrisation of a Jordan curve of the segmentation by its length. Assume that c is defined on some interval containing

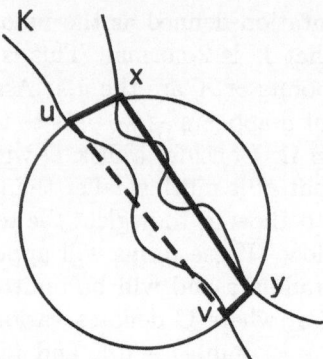

Figure 5.3 : No new crossing appears when
we pass to the limit.

$[-L, L]$. Then the length of the curve between $c(-s)$ and any other point $c(b)$ is $b + s$. Thus we have $|c(s) - c(-s)| \leq 2s$, the equality being true only if c is an affine curve. Set $|c(s) - c(-s)| = 2(s - \varepsilon)$. Notice that all points $c(r)$ with r in $[-s, s]$ must be enclosed by an ellipse C with foci $c(-s)$ and $c(s)$, defined by the relation: y is on C if and only if $|c(s) - y| + |y - c(-s)| = 2s$. Indeed, the total length of c is $2s$. One easily sees that the ellipse is contained in a rectangle, with sides of respective lengths $2s$ and, by Pythagoras relation, $2(s^2 - (s - \varepsilon)^2)^{\frac{1}{2}}$. Thus the area of the half rectangle is bounded from above by $2s(2s\varepsilon)^{\frac{1}{2}}$.

We shall now use the optimality of the segmentation with respect to the energy E by stating (like in the preceding lemma) that a certain rearrangement of the segmentation cannot decrease the energy. The rearrangement consists of replacing the curve c by the line segment $[c(-s), c(s)]$ and of course changing the value of u at points of the rectangle which are now contained in a new connected component. This rearrangement decreases the length term of E by $\lambda \varepsilon$ and increases the integral term of E by not more than half the area of the rectangle multiplied by $osc(g)^2$. Thus, by the optimality of the segmentation, $\lambda \varepsilon \leq 2s(2s\varepsilon)^{\frac{1}{2}} osc(g)^2$ and therefore $\varepsilon \leq 8s^3 . \frac{osc(g)^4}{\lambda^2}$.

We now consider two successive increments of c, $v_1 = c(0) - c(-s)$ and $v_2 = c(s) - c(0)$. We want to estimate the difference of these increments because it will yield to an estimate for the curvature at 0. We shall use the classical parallelogram identity, $|v_1 - v_2|^2 \leq 2(|v_1|^2 + |v_2|^2) - |v_1 + v_2|^2$. Thus $|v_1 - v_2|^2 \leq 4s^2 - (2s - 16s^3)^2 \leq 64s^4 \frac{osc(g)^4}{\lambda^2}$. Finally,

$$|v_1 - v_2| \leq 8s^2 \frac{osc(g)^2}{\lambda} .$$

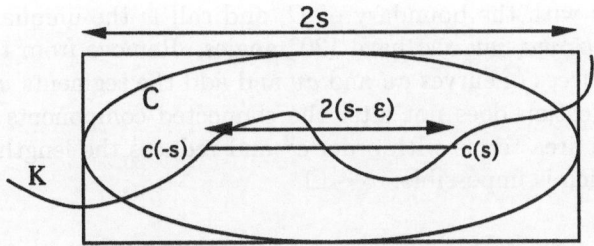

Figure 5.4 : estimate of the curvature of K.

This yields an upper bound for the (discretized) second-order derivative of c. In order to pass to a bound for the infinitesimal derivative, we use the fact that c is almost everywhere differentiable (which is proved in Chapter 11, Lemma 11.10), and we pick points a and b at which c is differentiable. We divide the interval $[a, b]$ in n equal intervals with vertices $s_i, i = 1, ..., n$. Set $w_i = \frac{n}{b-a}(c(s_i) - c(s_{i-1}))$. Then, using the above estimate with $s = \frac{b-a}{n}$, we get

$$|w_i - w_{i-1}| \leq 8\frac{b-a}{n} \cdot \frac{osc(g)^2}{\lambda}.$$

By adding all these estimates and using the triangular inequality,

$$|w_1 - w_n| \leq 8(b-a)\frac{osc(g)^2}{\lambda}.$$

Letting n tend to infinity, w_1 tends to $c'(a)$ and w_n tends to $c'(b)$, and therefore

$$c'(a) - c'(b) \leq 8(b-a)\frac{osc(g)^2}{\lambda}.$$

We conclude that c' is defined everywhere and Lipschitz with a constant less than $8\frac{osc(g)^2}{\lambda}$. Thus c'' is defined in the distributional sense and with modulus bounded from above by this constant. Now, the modulus of c'' is nothing but the curvature of c, and the proof is complete.   □

Lemma 5.9. *The crossing points of K are as announced in the main theorem of the introduction.*

Proof. Assume by contradiction that more than three curves arrive at a crossing point a or that exactly three arrive, but with angles different from 120°. Then two of these curves form an angle strictly less than 120°. Let ε be very small and D as above. Call u and v the first crossings of both curves with the boundary of D, and call w the unique point such that the lines wu, wv, wa have 120° angles. Remove from the segmentation the pieces of curves au and av and add the segments wu, wv, wa. This modification does not alter the connected components outside D, changes the area terms with order ε^2 and reduces the length of K with order ε, which is impossible. $\square$

We finish this section by collecting all the above results and proving the main existence and regularity theorem. There is only one thing to make more precise in Theorem 5.1: What is the set of segmentations over which we wish to minimize the Mumford and Shah functional? Clearly, if a segmentation has minimal energy, it must be (or can be made) 2-normal since the merging of two regions cannot decrease its energy. There is therefore no restriction in only minimizing the Mumford and Shah functional over all 2-normal segmentations. By Lemma 5.7, such minimal segmentations exist, and by Lemmas 5.8 and 5.9, every minimal 2-normal segmentation has the regularity properties announced in Theorem 5.1. Thus the proof of Theorem 5.1 is complete. The meaning of Theorem 5.1, however, is more theoretical than practical since there is no computational theory ensuring that an optimal segmentation can be computed in practice. Now, as we have emphasized, several of the lemmas leading to Theorem 5.1 are theoretical justifications for computing 2-normal, not necessarily optimal segmentations. The most relevant is Lemma 5.6, which ensures that the set of 2-normal segmentations is compact for the Hausdorff distance and therefore "small". So, unless we are not able to compute minimal segmentations, elements of a small set containing them are easily computed by a very simple computational tool: the region merging. As a consequence of this remark, we shall return to the region growing algorithms discussed in Chapter 3. Indeed, we can now interpret them as tools computing a mathematically sound class of segmentations: the 2-normal segmentations.

4. A pyramidal algorithm computing 2-normal segmentations.

We now consider the problem of computing a 2-normal segmentation. Notice that not all 2-normal segmentations are equally interesting: for instance, the empty segmentation, where Ω is the single region, clearly is a 2-normal segmentation. If the scale parameter λ is very large, it also is the right segmentation since by Lemma 5.5 the number α of regions is zero for λ large enough. However, it is obvious from the definition that the empty segmentation is 2-normal for every λ, which certainly proves that the assertion that a segmentation is 2-normal is not enough to ensure that it is "good". But if we follow the main idea of the region growing methods, as discussed in Chapter 3, *we see that what they compute is precisely a 2-normal subsegmentation of a fine initial segmentation, obtained by recursive merging.*

Assume that the datum g is defined on a rectangle. This rectangle is divided in small squares of constant size (the "pixels" or "picture elements"), and g is assumed to be constant on each pixel. Here are the properties which we require for the segmentations computed by a region growing algorithm, defined as an application associating with g and λ a segmentation (u, K).

a) "Correctedness" (fixed point property): Assume that g is piecewise constant on some areas of the rectangle. Then there exists a value λ_0 of the parameter λ such that for every $\lambda < \lambda_0$, the segmentation (u, K) obtained by the algorithm satisfies $u = g$ and K is the union of the boundaries of the regions where g is constant.

b) "Strong Causality" (pyramidal segmentation property): If $\lambda > \lambda'$, then the boundaries provided by the algorithm for λ are contained in those obtained for λ', and the regions of the segmentation associated to λ are the unions of some of the regions obtained for λ'. (This property has been discussed in Chapter 1.)

This last property ensures that a fast pyramidal algorithm can be implemented, computing a hierarchy of segmentations from fine to coarse scales. Moreover, the coarser segmentation will be deduced from the finer by "merging" operations, with a pyramidal structure for the computation. As a consequence of the "correctedness", if λ is very small, the computed segmentation is attained with (u_0, K_0) where $u_0 = u$ and K_0 consists of all the boundaries of all the pixels, and therefore coincides with the global minimum as λ is zero. We shall call the segmentation where each pixel is a region the "trivial segmentation". The recursive merging algorithm which we now define satisfies all the above-mentioned properties.

Algorithm computing a hierarchy of 2-normal segmentations.

Initialization. Set $u_0 = g$. u_0 is piecewise constant on the pixels, which correspond to small squares of a digital screen. K_0 is the union of all boundaries of all pixels. $\lambda_0 = 0$.

Recursive merging. Merge recursively all pairs of regions whose merging decreases the energy E.

Change of scale. Increase λ and return to the previous step.

5. Examples.

We now show two examples of segmentations obtained by Georges
Koepfler [KoeMS] and Christian Lopez [KLM]. In Figure 5.5, we present
in left-right and up-down order

  a) the original image g;
  b) its associated piecewise constant function u;
  c) the boundaries K of the constancy regions of u; and
  d) the superposition of K and g.

The choice of λ is *a priori* arbitrary. Now, the normalization of digital
pictures, which depend on technological and perceptual factors, led to
fix in the following examples the parameter λ between 2048 and 4096
for 512x512 pictures with 256 grey levels. Notice, however, that λ really
corresponds to a geometrical parameter of the segmentations if we trust
the estimates developed above. Indeed, we proved that the number of
regions behaves as $\frac{1}{\lambda^2}$, the length of an edge and its curvature as $\frac{1}{\lambda}$. It is
indeed the case in the numerical simulations.

In Figure 5.6, we present in the same format the segmentation of a
texture image. Here, the input g is a vector value function which associates with each point x several (18) "local features". In the case of the
experiment, the local features are simply gradients of smoothed versions
of the picture ("multiscale gradients") computed by a multiscale Haar
Wavelet transform. When the algorithm is imposed to stop when only
two regions are left, it yields the right texture regions.

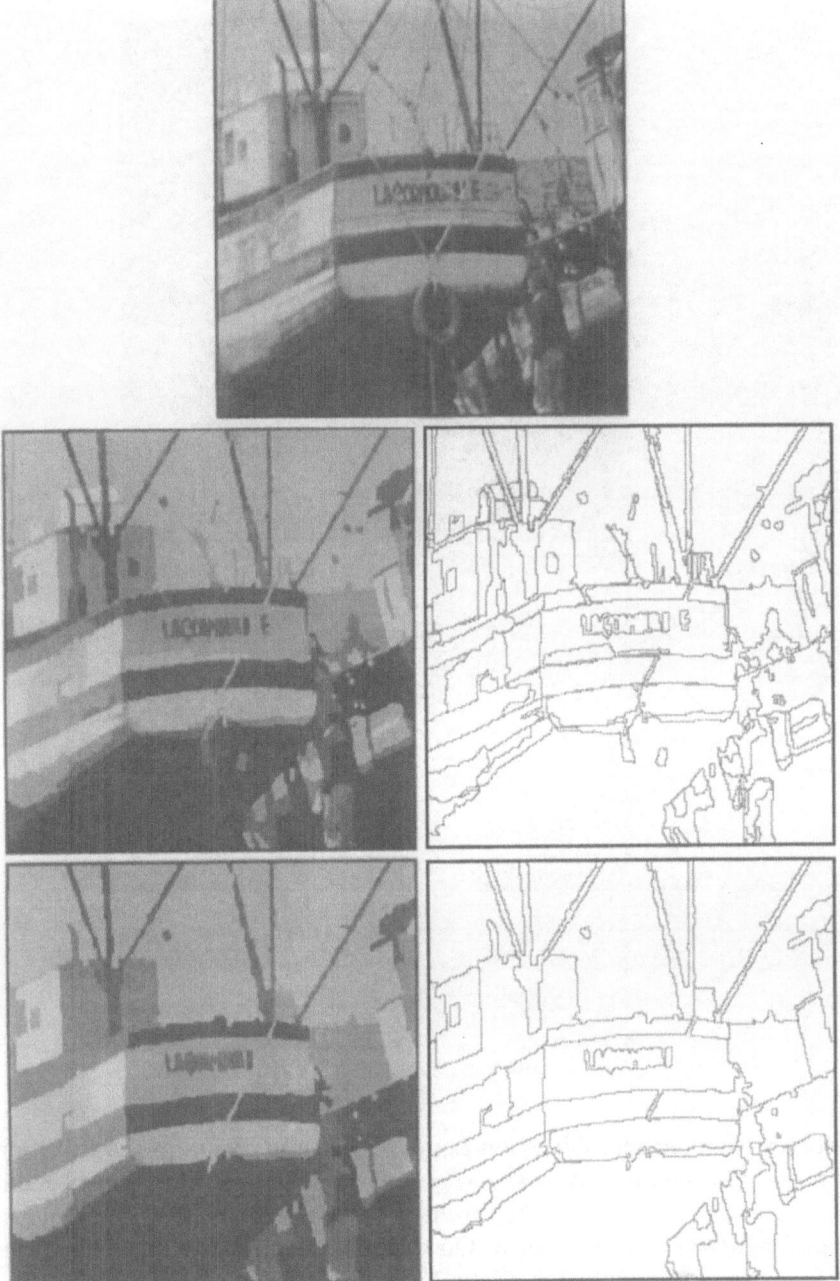

Georges Koepfler: Multiscale segmentation of a gray-level image (top) by the Mumford and Shah piecewise constant model at two scales (200 regions in the middle, 50 regions on the bottom). On the left u (piecewise constant), on the right K.

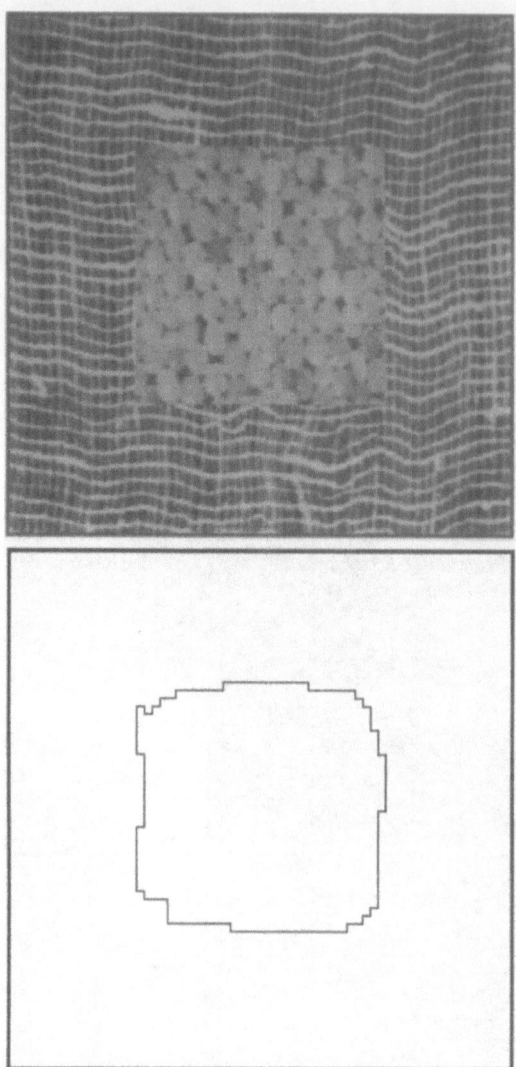

Georges Koepfler and Christian Lopez: Piecewise constant segmentation of a multichannel image by the Mumford and Shah model. The 18 channels (not shown) are computed from the original image (top) by a multiscale Haar (wavelet) transform. On the bottom are shown the boundaries of a 2-normal segmentation of the 18-channel image with 2 regions left.

Part II

ELEMENTS OF

GEOMETRIC MEASURE THEORY

Chapter 6

HAUSDORFF MEASURE

From now on, we abandon the heuristic style which was used in Part I. We shall give a rigorous definition of one of the terms used previously: the length. In this chapter, we define the Hausdorff outer measure of dimension α in a metric space E. We shall see that when $E = \mathbb{R}^N$ and α is an integer, this definition is a generalization of length ($\alpha = 1$), area ($\alpha = 2$), volume ($\alpha = 3$). Those quantities can be easily defined for smooth curves, surfaces, etc. The main difficulty of a more general definition, suitable for objects like the "edges" in an image, is maintaining the basic properties which are expected from such a measure, particularly the additivity. It must be emphasized that, in the very general framework of outer measures, the additivity is not ensured. In the following, we shall define the main formal properties of outer measures and give a characterization of the sets on which an outer measure becomes additive (measurable sets) when it is defined on a topological space (Section 2). Then, we shall apply this theory to the Hausdorff outer measure. Sections 3 and 4 are devoted to two basic examples of sets with finite Hausdorff length: the path connected sets on the one side which will be a paradigm for "edge sets". The second example, somehow orthogonal to the first, is the classical two-dimensional Cantor set. This set shows all characteristics of what we shall later call "fully irregular" or "fully unrectifiable" sets.

1. Hausdorff outer measure.

We consider a metric space E, that is, a set endowed with distance $d(x, y)$. By diam X we denote the supremum of the values $d(x, y)$ when x and y belong to X. In practice, the main example to keep in mind is the euclidean space $E = \mathbb{R}^N$. We denote by $\mathcal{A}, \mathcal{B}, \ldots$ sets of subsets of a metric space E. For $\alpha > 0$ let

$$|\mathcal{A}|_\alpha = \sum_{X \in \mathcal{A}} (\text{diam } X)^\alpha$$

$$\rho(\mathcal{A}) = \sup_{X \in \mathcal{A}} (\text{diam } X).$$

For a given $A \subset E$ we denote by $\mathcal{C}(A)$ the set of the countable coverings of A, namely the set of the sets $\mathcal{A}$ of subsets of E which are at most countable and such that

$$A \subset \bigcup \mathcal{A} .$$

(The last notation denotes the union of all the sets which belong to $\mathcal{A}$.) We define the *Hausdorff outer measure* $\mathcal{H}^\alpha(A)$ as

(6.1)
$$\mathcal{H}^\alpha(A) \;=\; c_\alpha \lim_{\substack{\mathcal{A} \in \mathcal{C}(A) \\ \rho(\mathcal{A}) \to 0}} \inf |\mathcal{A}|_\alpha ,$$

where c_α is a normalisation constant. The choice of the normalisation constant c_α is made in such a way that when α is an integer c_α is given by the Lebesgue measure of the unit ball in $\mathbb{R}^\alpha$ divided by 2^α. This choice makes $\mathcal{H}^\alpha$ agree with the Lebesgue outer measure on $\mathbb{R}^\alpha$, as we are going to see in the next chapter.

We can easily recognize that $\mathcal{H}^\alpha$ defines as an outer measure on E.

Outer measures. An *outer measure* on a set E is a function μ defined on the set $\mathcal{P}(E)$ of all subsets of E and valued in $[0, +\infty]$, which is *countably subadditive*, namely

(6.2)
$$\mu(\emptyset) \;=\; 0$$

(6.3)
$$\forall A \subset E , \; (A_n)_{n \in \mathbb{N}} \in \mathcal{P}(E)^{\mathbb{N}} \quad \text{such that} \quad A \subset \bigcup_n A_n :$$
$$\mu(A) \;\leq\; \sum_n \mu(A_n) .$$

Note that the above conditions imply the *finite subadditivity*, i.e.

(6.4) $\forall A, B, C \subset E$ such that $A \subset B \cup C :$ $\quad \mu(A) \leq \mu(B) + \mu(C)$

(the condition (6.2) is itself a condition of "zero subadditivity") and the *monotonicity* which means that

(6.5) $\qquad \forall A, B \subset E , \; A \subset B :$ $\quad \mu(A) \leq \mu(B) .$

We shall consider the set $\mathcal{M}_\mu$ of the subsets of E which are *measurable with respect to* μ *in the sense of Carathéodory*. A set A is in $\mathcal{M}_\mu$ by definition if

$$(6.6) \qquad \forall\, X \subset E : \quad \mu(X) \geq \mu(X \cap A) + \mu(X \setminus A) .$$

It is clear that, by (6.4), the above inequality is actually an equality. *A set is Carathéodory measurable if it divides every other set into two sets whose measures add.*

If $K \subset E$ we can consider the function μ_K from $\mathcal{P}(E)$ to $[0, +\infty]$ defined by

$$(6.7) \qquad \forall\, A \subset E : \quad \mu_K(A) = \mu(A \cap K) .$$

Then one sees immediately that μ_K still is an outer measure and that

$$(6.8) \qquad \mathcal{M}_\mu \subset \mathcal{M}_{\mu_K} .$$

We claim that $\mathcal{M}_\mu$ is a *σ-algebra* , namely that it satisfies the thesis of the following proposition.

Proposition 6.9. *For any outer measure μ on a set E we have*

$$(6.10) \qquad\qquad \emptyset \in \mathcal{M}_\mu$$

$$(6.11) \qquad \forall\, A \in \mathcal{M}_\mu : \quad E \setminus A \in \mathcal{M}_\mu$$

$$(6.12) \qquad \forall\, (A_n)_{n \in \mathbb{N}} \in (\mathcal{M}_\mu)^{\mathbb{N}} : \quad \bigcup_n A_n \in \mathcal{M}_\mu .$$

(6.10) and (6.11) are trivial. For the proof of (6.12), which will be given later, we first point out the corresponding property for a finite union.

Lemma 6.13. *If μ is an outer measure on a set E, then*

$$(6.14) \qquad \forall A, B \in \mathcal{M}_\mu : \quad A \cup B \in \mathcal{M}_\mu .$$

Proof. Fix X in $\mathcal{P}(E)$. Note that

$$X \setminus (A \cup B) = (X \setminus A) \setminus B$$

and
$$X \cap (A \cup B) = (X \cap A) \cup ((X \setminus A) \cap B).$$

Therefore since A and B are in $\mathcal{M}_\mu$, we have

$$\begin{aligned}
\mu(X \cap (A \cup B)) &+ \mu(X \setminus (A \cup B)) \\
&\leq \mu(X \cap A) + \mu((X \setminus A) \cap B) + \mu((X \setminus A) \setminus B) \\
&= \mu(X \cap A) + \mu(X \setminus A) = \mu(X),
\end{aligned}$$

where we have used the fact that A and B enjoy (6.6) in the last two steps. So $A \cup B$ also enjoys (6.6) for every choice of X and therefore it belongs to $\mathcal{M}_\mu$. $\quad\square$

From (6.11) and (6.14) one deduces the finite intersection property of $\mathcal{M}_\mu$

$$(6.15) \qquad\qquad \forall\, A, B \in \mathcal{M}_\mu\,, \qquad A \cap B \in \mathcal{M}_\mu$$

and from (6.6) (taking $X = A \cup B$) follows the finite additivity of μ on $\mathcal{M}_\mu$

$$(6.16) \qquad \forall\, A, B \in \mathcal{M}_\mu\,, \ A \cap B = \emptyset : \quad \mu(A \cup B) = \mu(A) + \mu(B).$$

(Actually, the fact that at least one among A and B is in $\mathcal{M}_\mu$ is clearly enough in order to ensure (6.16).) The above formula can easily be extended to give the *countable additivity* of an outer measure on the measurable sets, as stated in the following proposition.

Proposition 6.17. *If $(A_n)_{n \in \mathbb{N}}$ is a sequence of disjoint sets in $\mathcal{M}_\mu$, then*

$$(6.18) \qquad\qquad \mu\left(\bigcup_n A_n \right) = \sum_n \mu(A_n).$$

Proof. One inequality trivially follows from (6.3). For the other one we see that by iterating (6.16) we have for every value of n

$$\sum_{k=0}^{n} \mu(A_k) = \mu\left(\bigcup_{k=0}^{n} A_k \right) \leq \mu\left(\bigcup_k A_k \right)$$

where the last inequality follows by (6.5). So we obtain the other inequality in (6.18). $\quad\square$

Proposition 6.19. *If* $(A_n)_{n \in \mathbb{N}} \in (\mathcal{M}_\mu)^{\mathbb{N}}$, *then*

$$(6.20) \qquad \mu \left(\bigcup_n A_n \right) = \lim_n \mu \left(\bigcup_{k=0}^{n} A_k \right).$$

Proof. Let for every n in $\mathbb{N}$

$$B_n = A_n \setminus \bigcup_{k=0}^{n-1} A_k.$$

From (6.16), since the B_n are disjoint,

$$\mu \left(\bigcup_{k=0}^{n} A_k \right) = \mu \left(\bigcup_{k=0}^{n} B_k \right) = \sum_{k=0}^{n} \mu(B_k)$$

and from (6.18)

$$\mu \left(\bigcup_n A_n \right) = \mu \left(\bigcup_n B_n \right) = \sum_n \mu(B_n).$$

Therefore, (6.20) is equivalent to

$$\sum_n \mu(B_n) = \lim_n \left(\sum_{k=0}^{n} \mu(B_k) \right)$$

which is obviously true. □

A different formulation of (6.20) is
(6.21)

$$\forall (A_n)_{n \in \mathbb{N}} \in (\mathcal{M}_\mu)^{\mathbb{N}}, \ A_n \subset A_{n+1} : \quad \mu \left(\bigcup_n A_n \right) = \lim_n \mu(A_n).$$

Remark. In the above equality, the measurability assumption on the sets A_n can be relaxed by asking for the property that every A_n is contained in a measurable set A'_n with the same outer measure. Indeed, in this case, one can always assume that the sets A'_n also form an increasing sequence by inclusion, after replacing every A'_n with the set $A''_n = \cap_{k \geq n} A'_k$. Then

$$\mu(\cup_n A_n) \leq \mu(\cup_n A''_n) = lim_n \mu(A''_n) = lim_n \mu(A_n).$$

The possibility of including any set in a measurable set with the same outer measure is always ensured, as we shall see in the next section, when μ is a Hausdorff measure in a metric space.

By the difference from (6.21), with the help of (6.16) we also have

$$\forall \, (A_n)_{n \in \mathbb{N}} \in \mathcal{M}_\mu^{\mathbb{N}} \,, \ \mu(A_1) < +\infty \,, \ A_{n+1} \subset A_n \ :$$

(6.22)

$$\mu\left(\bigcap_n A_n\right) = \lim_n \mu(A_n) \,.$$

We are now in a position to finish the proof of Proposition 6.9.

Proof of (6.12). We know from Lemma 6.13 that, for every value of n, $\bigcup_{k=0}^n A_k \in \mathcal{M}_\mu$ and therefore for every $X \subset E$,

$$\mu(X) = \mu\left(X \cap \bigcup_{k=0}^n A_k\right) + \mu\left(X \setminus \bigcup_{k=0}^n A_k\right)$$

$$\geq \mu\left(X \cap \bigcup_{k=0}^n A_k\right) + \mu\left(X \setminus \bigcup_k A_k\right).$$

So we apply (6.20) to the outer measure μ_X and, since by (6.8) the sets A_k are in $\mathcal{M}_{\mu_X}$, we obtain

$$\mu(X) \geq \mu\left(X \cap \bigcup_k A_k\right) + \mu\left(X \setminus \bigcup_k A_k\right)$$

which means, since X is arbitrary, that $\bigcup_k A_k \in \mathcal{M}_\mu$. $\square$

We have proved in this way that $\mathcal{M}_\mu$ is a σ-algebra and that (the restriction of) μ is a measure on $\mathcal{M}_\mu$. In the concrete cases one is interested in characterizing the sets in $\mathcal{M}_\mu$ or in its subset $\mathcal{I}_\mu$ consisting of the sets of $\mathcal{M}_\mu$ at which μ takes a finite value. $\mathcal{I}_\mu$ clearly contains the set $\mathcal{I}_\mu^0$ consisting of all the subsets of E at which μ takes value 0 (*the negligible sets*), since in correspondence with those sets, (6.6) is satisfied as a consequence of (6.5).

2. Outer measures in topological spaces.

When E is a topological space, one can compare $\mathcal{M}_\mu$ with the σ-algebra $\mathcal{B}$ of the *Borel sets*, defined as the *smallest σ-algebra containing all the open sets of E*. For the case of a metric space one can take advantage of the following criterion.

Proposition 6.23 (Carathéodory criterion). *If E is a metric space then $\mathcal{B} \subset \mathcal{M}_\mu$ if and only if the additivity relation (6.16) is satisfied by any pair of sets (A, B) such that $d(A, B) > 0$.*

Proof. If $d(A, B) > 0$ then

$$A = (A \cup B) \cap \overline{A}$$
$$B = (A \cup B) \setminus \overline{A}.$$

Now, if $\mathcal{B} \subset \mathcal{M}_\mu$ then $\overline{A} \in \mathcal{M}_\mu$ and the additivity relation (6.16) $\mu(A \cup B) = \mu(A) + \mu(B)$ follows from (6.6).

Conversely, let A be an open set and fix any $X \in \mathcal{P}(E)$. Let, for $k \in \mathbb{N} \setminus \{0\}$,

$$A_k = \left\{ x \in A \mid d(x, E \setminus A) > \frac{1}{k} \right\},$$

$$B_k = A_{k+1} \setminus A_k.$$

Then if $k \geq h + 2$ one has $d(B_k, B_h) \geq \frac{1}{k-1} - \frac{1}{k} > 0$. So we can apply the additivity relation (6.16) to the sets $X \cap B_k$ and $X \cap B_h$. By iterating this argument and by monotonicity, we have for any $m \in \mathbb{N}$

$$\sum_{i=1}^{m} \mu(X \cap B_{2i}) = \mu\left(\bigcup_{i=0}^{m} X \cap B_{2i}\right) \leq \mu(X)$$

and therefore

$$\sum_{k \text{ even}} \mu(X \cap B_k) \leq \mu(X).$$

In the same way one finds the same bound on the sum extended to odd values of k and obtains

$$\sum_{k} \mu(X \cup B_k) \leq 2\mu(X).$$

So, if $\mu(X)$ is finite,

$$\lim_{k} \sum_{i \geq k} \mu(X \cap B_i) = 0$$

and, subsequently, by the subadditivity of μ,

$$\lim_{k} \mu(X \cap (A \setminus A_k)) = \lim_{k} \mu\left(X \cap \bigcup_{i \geq k} B_i\right)$$

$$\leq \lim_{k} \sum_{i \geq k} \mu(X \cup B_i) = 0.$$

(A being open, one has $A \setminus A_k = \bigcup_{i \geq k} B_i$.) Now, since $X \cap A = (X \cap A_k) \cup (X \cap (A \setminus A_k))$, we get by subadditivity $\mu(X \cap A) \leq \mu(X \cap A_k) + \mu(X \cap (A \setminus A_k))$, and therefore $\mu(X \cap A) \leq \lim_k \mu(X \cap A_k)$. The monotonicity of μ yields the converse inequality and we obtain

$$(6.24) \qquad \mu(X \cap A) = \lim_k \mu(X \cap A_k) \, .$$

For any value of k we have $d(A_k, X \setminus A) \geq \frac{1}{k} > 0$; hence we can apply the additivity relation (6.16) and we have

$$\mu(X) \geq \mu((X \cap A_k) \cup (X \setminus A)) = \mu(X \cap A_k) + \mu(X \setminus A) \, .$$

Then we pass to the limit by (6.24) and obtain (6.6), which means that $A \in \mathcal{M}_\mu$. $\qquad \square$

Application to Hausdorff measures.

We can apply the previous criterion to the case in which μ is the Hausdorff measure of a given dimension α. Let A and B be two subsets of E such that $d(A, B) > 0$. Then if $\mathcal{A} \in \mathcal{C}(A \cup B)$ is a covering of $A \cup B$ and if $\rho(\mathcal{A}) < d(A, B)$, we can write $\mathcal{A} = \mathcal{A}_1 \cup \mathcal{A}_2$ with $\mathcal{A}_1 \cap \mathcal{A}_2 = \emptyset$ and $\mathcal{A}_1 \in \mathcal{C}(A)$, $\mathcal{A}_2 \in \mathcal{C}(B)$. Therefore

$$|\mathcal{A}|_\alpha = |\mathcal{A}_1|_\alpha + |\mathcal{A}_2|_\alpha \, .$$

We can let $\rho(\mathcal{A}) \to 0$ and $c_\alpha |\mathcal{A}|_\alpha \to \mathcal{H}^\alpha(A \cup B)$. This yields

$$\mathcal{H}^\alpha(A \cup B) \geq \mathcal{H}^\alpha(A) + \mathcal{H}^\alpha(B) \, .$$

So $\mathcal{H}^\alpha$ satisfies the Carathéodory criterion and therefore the Borel sets are measurable with respect to $\mathcal{H}^\alpha$.

Since every set is contained in a closed set with the same diameter and in an open set with a diameter arbitrarily close to the previous one, then one can assume without any restriction that *the limit in the relation defining the Hausdorff measure (6.1) can be taken only on open coverings or only on closed coverings* $\mathcal{A}$. Therefore, given any subset A of E we can fix a sequence of open coverings $\mathcal{A}_n$ such that

$$(6.25) \qquad \rho(\mathcal{A}_n) \to 0, \qquad c_\alpha |\mathcal{A}_n|_\alpha \to \mathcal{H}^\alpha(A) \, .$$

If $A_n = \bigcup \mathcal{A}_n$ and $A' = \bigcap_n A_n$, we see that A' is a G_δ-set (that is, a countable intersection of open sets and therefore a Borel set) with

$A \subset A'$ and that for every n, $A_n \in C(A')$. Therefore, from (6.25) and by monotonicity, $\mathcal{H}^\alpha(A') = \mathcal{H}^\alpha(A)$.

So, if A is a $\mathcal{H}^\alpha$-measurable set such that $\mathcal{H}^\alpha(A) < +\infty$, we can write

$$\mathcal{H}^\alpha(A') = \mathcal{H}^\alpha(A' \setminus A) + \mathcal{H}^\alpha(A)$$

and therefore $\mathcal{H}^\alpha(A' \setminus A) = 0$. Thus A is the difference set between a Borel set A' and a negligible set $A' \setminus A$. On the other hand, since by the previous analysis itself, $A' \setminus A$ is contained in a Borel negligible set B, then A also is the union of the Borel set $A' \setminus B$ and the negligible set $B \setminus (A' \setminus A)$. In conclusion we see that *the measurable sets of finite measure defined by $\mathcal{H}^\alpha$ are precisely the sets obtained by adding or subtracting a negligible set to a Borel set of finite outer measure.* The last conclusion can be sharpened in the following way.

Proposition 6.26. *If A is a measurable set with finite $\mathcal{H}^\alpha$-measure, then*

$$(6.27) \qquad \mathcal{H}^\alpha(A) = \sup_{\substack{C \subset A \\ C \text{ closed}}} \mathcal{H}^\alpha(C).$$

Proof. Fix any $a < \mathcal{H}^\alpha(A)$ and let b be a number such that $a < b < \mathcal{H}^\alpha(A)$. Let $(A_n)_{n \in \mathbb{N}}$ be a sequence in $C(A)$ such that (6.25) holds. As we noticed above, we may assume that all the elements in A_n are closed. Let A_n be the union of all the elements of A_n and A'_n be a finite union of elements of A_n recursively chosen in such a way that

$$(6.28) \qquad \mathcal{H}^\alpha \left(A \cap \bigcap_{k=1}^{n} A'_k \right) \geq b.$$

This is possible by the countable additivity (6.20) of the Hausdorff measure, since the left-hand side of (6.28) is $\mathcal{H}^\alpha(A)$ if we replace every A'_k by A_k.

Then if $A' = \bigcap_{n=1}^{\infty} A'_n$ and $A'' = \bigcap_{n=1}^{\infty} A_n$, we see that A' is closed, $A' \subset A''$, $A \subset A''$ and, by (6.22) ,

$$(6.29) \qquad \mathcal{H}^\alpha(A \cap A') \geq b.$$

By (6.1) and (6.25)

$$(6.30) \qquad \mathcal{H}^\alpha(A'') \leq \lim_n c_\alpha |A_n|_\alpha \leq \mathcal{H}^\alpha(A).$$

So we have that $A \subset A''$, $\mathcal{H}^\alpha(A) = \mathcal{H}^\alpha(A'')$ and

$$\mathcal{H}^\alpha(A' \setminus A) \leq \mathcal{H}^\alpha(A'' \setminus A) = 0 \, .$$

Thus, by the remarks following (6.25), $A' \setminus A$ is contained in a G_δ negligible set G. Let $G = \bigcap_{n=1}^\infty B_n$, with every B_n open. We have

$$\bigcup_n (A' \setminus B_n) = A' \setminus G \subset A' \setminus (A' \setminus A) \subset A \, .$$

Moreover, by (6.20) and (6.28)

$$\mathcal{H}^\alpha(A' \setminus B_n) \to \mathcal{H}^\alpha(A' \setminus G) = \mathcal{H}^\alpha(A') \geq b \, .$$

So we can find n such that (the closed subset of A) $A' \setminus B_n$ has a measure larger than a. □

We end this section with a useful property of coverings of negligible sets: Every $\mathcal{H}^\alpha$-negligible set in a metric space has coverings for which the sum of the diameters to the power α is arbitrarily small and such that every point of the set belongs to a suitably small element of the covering.

Lemma 6.31. *Let X be a subset of E, ρ be a function defined on X which takes strictly positive values, and α be a positive number such that $\mathcal{H}^\alpha(X) = 0$. Then, for every $\varepsilon \geq 0$, one can find a covering $\mathcal{B}$ such that every B in $\mathcal{B}$ is a ball centered at some point x of X, with a radius smaller than $\rho(x)$ and such that $|\mathcal{B}|_\alpha \leq \varepsilon$.*

Proof. For every integer $n \geq 1$, we consider the subset X_n of X consisting of the points x such that $\rho(x) \geq \frac{1}{n}$. Since, for every n, X_n is a $\mathcal{H}^\alpha$-negligible set, we can find a covering $\mathcal{A}_n$ of X_n such that $\rho(\mathcal{A}_n) \leq \frac{1}{n}$ and $|\mathcal{A}_n|_\alpha \leq \frac{\varepsilon}{2^{n+\alpha}}$. Moreover, we can assume that every A in $\mathcal{A}_n$ contains an element x_A of X_n. For every A, we consider the ball B_A centred at x_A and with radius $\text{diam}(A)$. We call $\mathcal{B}_n$ the set of such balls obtained for A in $\mathcal{A}_n$. Then $\mathcal{B}_n$ is a covering of X_n and $|\mathcal{B}_n|_\alpha \leq \frac{\varepsilon}{2^n}$. One easily sees that the set $\mathcal{B} = \bigcup_{n \geq 1} \mathcal{B}_n$ is a covering satisfying the required properties.

□

3. Connected sets with finite one-dimensional Hausdorff measure.

An immediate consequence of the definition of Hausdorff measure is that if f is a Lipschitz map between two given metric spaces, with Lipschitz constant L, then for every set A contained in the space of definition,

$$(6.32) \qquad \mathcal{H}^\alpha(f(A)) \leq L^\alpha \mathcal{H}^\alpha(A).$$

This result implies that *if a connected subset of $\mathbb{R}^n$ contains two points x and y, it has a $\mathcal{H}^1$-measure larger or equal to the distance between both points.* Indeed, the projection of the set on the straight line passing by x and y must contain the segment between x and y which has a $\mathcal{H}^1$-measure equal to the distance between x and y. (This last point is easy to check, and will anyway be proved at the end of the next chapter.) On the other hand, the projection has a Lipschitz constant equal to 1; so the measure of the segment must be smaller or equal to the measure of the set. As another direct consequence of the definition of the Hausdorff measure, we shall now give a characterization of the compact connected subsets of $\mathbb{R}^n$ with finite $\mathcal{H}^1$-measure. We say that a set T is *path connected* if every two points in it can be connected by a rectifiable curve.

Proposition 6.33. *Let T be a closed connected subset of $\mathbb{R}^n$ with finite $\mathcal{H}^1$-measure, then T is path connected.*

Proof. Let ε be a positive real number. By the definition of Hausdorff measure (6.1) we can find a countable covering $\mathcal{A}$ of T such that $\rho(A) \leq \varepsilon$ and $|\mathcal{A}|_1 \leq \mathcal{H}^1(T) + 1$. Using the remark following (6.25), we can assume that every set A in $\mathcal{A}$ is open and connected. Of course, we can also impose that every set in $\mathcal{A}$ meets T. Thus if N is the union of the sets in $\mathcal{A}$, it is connected and, being open, path connected. So, fixing x and y in T, we can find a rectifiable curve contained in N which joins x and y. This curve is compact and therefore covered by a finite subset $\mathcal{B}$ of $\mathcal{A}$.

We denote by A_1 one of such sets which contains x and by x_1 the point where the curve definitively leaves A_1. In the same way, we call A_2 a set of $\mathcal{B}$ which contains x_1, by x_2 the last point where the curve meets the closure of A_2, and so on. After a finite number k of steps we must reach a set A_k containing y and we set $x_k = y$. Therefore x and y can be connected by the piecewise linear path c_ε with vertices x_i and length $\sum_{i=2}^{k} d(x_i, x_{i-1})$. This length is bounded from above by $\sum_{i=1}^{k} \mathrm{diam}(A_i)$ and, since the sets A_i are all disjoint by construction, by $|\mathcal{A}|_1 \leq \mathcal{H}^1(T) + 1$. We now let ε tend to zero. Each curve c_ε connecting x and y clearly is contained in the 2ε-neighbourhood of T. Since the c_ε have a uniformly bounded length, we can find a converging subsequence by the Ascoli-Arzela theorem and we obtain a limit rectifiable curve, contained in T because T is closed, which connects x and y. $\square$

4. A suggesting example of Hausdorff measurable set with finite length.

We shall end this chapter with a classical example which shows how the Hausdorff measure permits to define a "length" for sets which can in no way be considered as curves or subsets of curves. This example is important because it will also be referred to as a classical example of "nonrectifiable" or "irregular" set. We shall develop the theory of rectifiability and regularity in the next chapters. The set B will be defined as the limit of a sequence of sets obtained by iterating a contracting multivalued map $\mathcal{B}$ of the plane. Set for (x, y) in the plane

$$\mathcal{B}_1(x, y) = \frac{1}{3} \cdot (x, y)$$

$$\mathcal{B}_2(x, y) = \frac{1}{3} \cdot (x + 1, y + 1)$$

$$\mathcal{B}_3(x, y) = \frac{1}{3} \cdot (x + 2, y).$$

The multivalued map $\mathcal{B}$ is a contracting map for the Hausdorff distance and it admits, by the Cauchy-Lipschitz theorem, a unique closed attractor B defined $B = \lim_{n \to \infty} \mathcal{B}^n(z)$, where the limit is taken in the Hausdorff sense and does not depend on the starting point z. (We denote by $\mathcal{B}^n$ the n-times iteration of $\mathcal{B}$. The Hausdorff distance is defined in Lemma 5.6 and, more extensively, in the begining of Chapter 10.) In order to make things intuitive, we can iteratively construct B by a classical process due to Cantor. Let us start, instead of z, with the initial set B_0 defined as the segment S_0^1 of the plane with endpoints $(0, 0)$ and $(1, 0)$. We construct B_1 by dividing B_0 into three parts with equal length and lifting the middle part so that its height is equal to its length. The same process is iterated on each one of the three segments thus obtained. So the intersection of B with the line $\{y = 0\}$ is nothing but the classical one-dimensional Cantor set.

The set $B_n = \mathcal{B}^n(B_0)$ is made of 3^n horizontal segments S_n^i with length $\frac{1}{3^n}$. It is easily seen that $d(B_n, B) \leq 3^{-n}$. On Figure 6.1, we have represented $B_3 = \mathcal{B}^3(B_0)$.

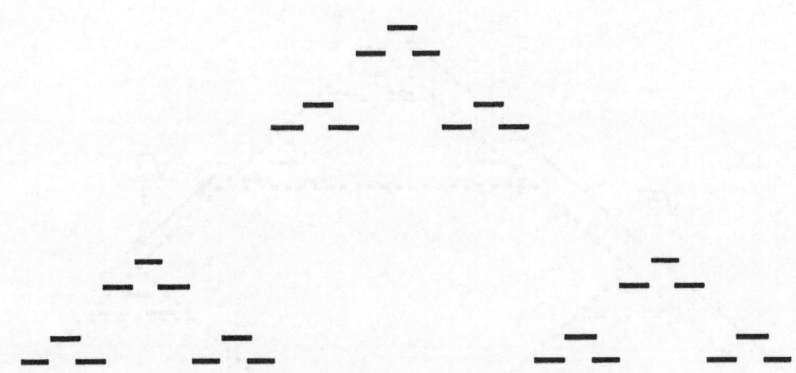

<u>Figure 6.1</u> : the set B3, which approximates the nonrectifiable set B.

Proposition 6.34. *The set B is a measurable set with Hausdorff length equal to 1.*

Proof. We shall use the definition of Hausdorff measure (6.1) and define a sequence of coverings whose diameter tends to zero and whose sum of diameters does not exceed 1. We simply remark that B_n is contained in a set $\mathcal{T}_n$ defined as the union of all rectangle isosceles triangles whose hypothenuses are the segments S_n^i ($1 \leq i \leq 3^n$). Denote by T_n^i the triangle corresponding to the segment S_n^i. In Figure 6.2, one can see the unique triangle of $\mathcal{T}_0$, the three triangles of $\mathcal{T}_1$, and the nine triangles of $\mathcal{T}_2$.

One easily checks the following properties:

1) For fixed n, $T_n^i \cap T_n^j = \emptyset$.
2) If $p > n$ then every triangle T_p^i is contained in a triangle T_n^j.
3) For every n, $B \subset \mathcal{T}_n$.
4) For every $n \geq p$, $\mathrm{diam}(T_p^i) = \frac{1}{3^p}$.

For every $\rho > 0$, we can find n such that $3^{-n} < \rho$. Thus a covering of B with sets of diameter less than ρ is given by the set $\mathcal{T}_n$ of the triangles T_n^i, and therefore

$$\mathcal{H}^1(B) \leq \liminf_n |\mathcal{T}_n|_1 \leq 1.$$

In order to get the converse inequality, we notice that the vertical projection of B onto the line $\{y = 0\}$ is B_0 and has therefore Hausdorff length 1. Since the projection is a contracting map, the converse inequality follows from (6.32). $\square$

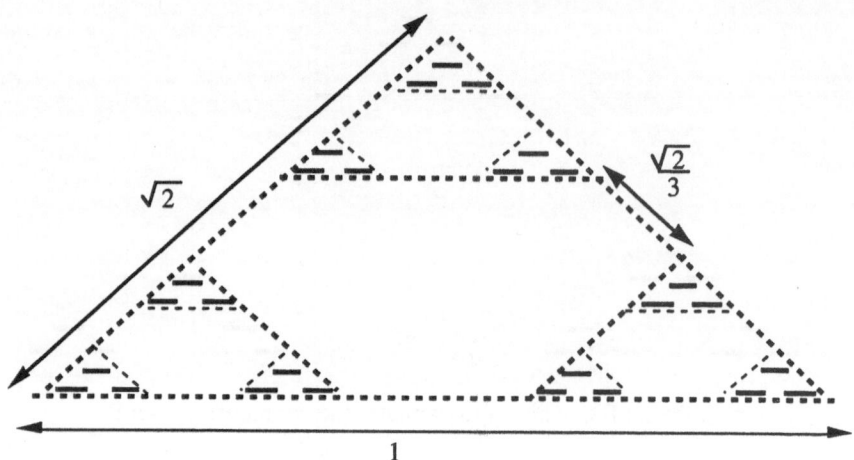

Figure 6.2 : the set B3, and the coverings T1 and T2 of B and B3.

In the following sections, we shall refer to the set B as the "planar Cantor set". We shall now prove a remarkable property of B, which indicates how much B differs from any curve or union of curves.

Proposition 6.36. *Denote by P_1 the vertical projection onto the line $\{x = 0\}$, and by P_2 the projection onto the line $\{x = y\}$. Then $\mathcal{H}^1(P_1 B) = \mathcal{H}^1(P_2 B) = 0$.*

Proof. It is easily seen that $\mathcal{H}^1(P_1 B) \leq \mathcal{H}^1(P_1 \mathcal{T}_n) = \frac{1}{2} \cdot \left(\frac{2}{3}\right)^n$ for every n and therefore $\mathcal{H}^1(P_1 B) = 0$.

In the same way, $\mathcal{H}^1(P_2(B)) \leq \mathcal{H}^1(P_2(\mathcal{T}_n)) = \frac{1}{\sqrt{2}} \cdot \left(\frac{2}{3}\right)^n$, which yields $\mathcal{H}^1(P_2(B)) = 0$. The same argument easily yields that the orthogonal projection of B onto every line except $\{y = 0\}$ has zero measure.

As we shall see in Chapters 8 to 12, the property that B has zero projection on more than one direction means that B is totally nonrectifiable, and if so, then it has a projection with zero measure onto all directions except a negligible subset of $[0, \pi]$. Let us limit ourselves for now to notice how much B differs from any curve: indeed, any curve which is not reduced to a single point has a projection with positive length onto every line but (at most) one.

Chapter 7

COVERING LEMMAS IN A METRIC SPACE

Of course, we only have access to the Hausdorff measure of a given set A by looking at its coverings. Clearly, the definition (6.1) is very abstract and gives no indication about how a covering of A should be to give an accurate account of the Hausdorff measure. This section is devoted to several criteria which will tell us when a covering is a "good" covering, correctly approximating the Hausdorff measure. As we shall see, and this is rather surprising, the first criterion simply is that the covering is made of sets with diameter small enough. We shall also focus on additional properties that we would like to require from a covering. The main property is that the sets of the covering should be disjoint. Indeed, if they are, we may add inequalities concerning A obtained on each one of the sets of the covering and obtain global estimates on A. Thus, we may pass from local estimates on a covering of A to global estimates on the set A itself. The Vitali Covering Lemma is an essential tool to do that and we shall use it constantly in the next chapters. This chapter ends with a classical application of the Vitali Covering Lemma: Lebesgue measure and the N-dimensional Hausdorff measure agree on $\mathbb{R}^N$.

1. Elementary properties of generic coverings.

If B is a set in a metric space E, we shall denote by B^* the 2 diam (B)-neighbourhood of B. For instance, if B is an open ball of $\mathbb{R}^N$, then B^* is the ball with the same center and with a radius five times larger. When $\mathcal{B}$ is a set of subsets of E, we shall denote by $\mathcal{B}^*$ the set of all the subsets B^* obtained when B is in $\mathcal{B}$.

Covering Lemma 7.1. *If $\mathcal{B}$ is any set of subsets of E such that $\rho(\mathcal{B}) < +\infty$, it contains a subset $\mathcal{B}'$ consisting of disjoint sets such that*

$$(7.2) \qquad \forall\, B \in \mathcal{B}\,,\ \exists\, B' \in \mathcal{B}'\quad such\ that\quad B \subset (B')^*$$

and subsequently

$$(7.3) \qquad\qquad \bigcup \mathcal{B} \subset \bigcup (B')^*\,.$$

Proof. Set $\rho(\mathcal{B}) = \bar{r}$. Define recursively a sequence of subsets $\mathcal{B}_n$ of $\mathcal{B}$ by setting $\mathcal{B}_0 = \emptyset$ and by taking for $\mathcal{B}_{n+1}$ a maximal set containing $\mathcal{B}_n$ and made of disjoint sets in $\mathcal{B}$ with diameter larger or equal to $\bar{r}2^{-n}$.

Let $\mathcal{B}'' = \bigcup_n \mathcal{B}_n$ and let $\mathcal{B}'$ be equal to $\mathcal{B}''$ plus the one-point sets in $\mathcal{B}$ which are not contained in any set in $\mathcal{B}''$. Now take $B \in \mathcal{B}$, and let r be its diameter. If $r = 0$ then either B is in $\mathcal{B}'$ or it is contained in a set belonging to $\mathcal{B}''$. If $r > 0$, fix $n \in \mathbb{N}$ such that

$$\bar{r}2^{-(n+1)} \leq r \leq \bar{r}2^{-n} .$$

If $B \notin \mathcal{B}_{n+2}$ then we can find, by the maximality property of $\mathcal{B}_{n+2}$, a set B' in $\mathcal{B}_{n+2}$ which has a nonempty intersection with B. By definition B' has a diameter r' larger or equal to $\bar{r}2^{-(n+1)}$ and therefore such that $r < 2r'$. Then (7.2) is obvious. $\square$

Lemma 7.4. *Let A be a subset of E such that $\mathcal{H}^\alpha(A) < +\infty$. Then in correspondence with any positive number ε one can find a number $\bar{\rho} > 0$ such that*

for every countable subset $\mathcal{A}$ of $\mathcal{P}(E)$ such that $\rho(\mathcal{A}) \leq \bar{\rho}$

$$\mathcal{H}^\alpha(A \cap \bigcup \mathcal{A}) \leq c_\alpha |\mathcal{A}|_\alpha + \varepsilon .$$
(7.5)

This is an important approximation lemma. It can be compared with its equivalent when A is a smooth curve: then the curve can be approximated by a sequence of segments (a piecewise affine curve) with endpoints on A. Provided the distance of consecutive vertices of this approximating curve is small enough, we know that the length of any subcurve C of A is close to the length of the approximating piecewise affine curve obtained by joining the segments with endpoints on C. (See beginning of Chapter 5.) This approximation uniformly holds with respect to C. Lemma 7.4 is a similar statement for the general Hausdorff measure. We can consider a covering of A as an approximating set whose length is evaluated in a simpler way by summing the diameters of the sets of the covering. Then the maximal diameter of the sets of the covering is the only parameter to take into account for knowing how good this evaluation is.

Proof. Take any $\mathcal{A}' \in \mathcal{C}(A \setminus \bigcup \mathcal{A})$ with $\rho(\mathcal{A}') \leq \bar{\rho}$. Then $\mathcal{A} \cup \mathcal{A}' \in \mathcal{C}(A)$ and $\rho(\mathcal{A} \cup \mathcal{A}') \leq \bar{\rho}$. So, by the definition of the Hausdorff measure, we can fix $\bar{\rho}$ in such a way that

$$\mathcal{H}^\alpha(A) - \varepsilon \leq c_\alpha |\mathcal{A} \cup \mathcal{A}'|_\alpha \leq c_\alpha \left(|\mathcal{A}|_\alpha + |\mathcal{A}'|_\alpha \right) .$$
(7.6)

By (6.1) we can let $c_\alpha |\mathcal{A}'|_\alpha \to \mathcal{H}^\alpha(A \setminus \bigcup \mathcal{A})$. So (7.6) becomes

$$\mathcal{H}^\alpha(A) \leq c_\alpha |\mathcal{A}|_\alpha + \varepsilon + \mathcal{H}^\alpha(A \setminus \bigcup \mathcal{A}) .$$
(7.7)

Now we just have to observe that there is no restriction to assume that every set belonging to $\mathcal{A}$ is closed (because, as observed in (6.25), the diameter does not increase if we replace a set by its closure), so $\bigcup \mathcal{A}$ is measurable. Therefore (7.5) is an obvious consequence of (7.7) and the additivity of the Hausdorff measure (6.6). $\square$

Remark. Note that the value of $\bar{\rho}$ determined in the previous proposition only depends on A and not on $A \cap \left(\bigcup \mathcal{A} \right)$. We shall say that $\mathcal{A}$ is an *almost covering* of A if $\mathcal{H}^{\alpha} \left(A \setminus \bigcup \mathcal{A} \right) = 0$. The above statement tells in particular that the "lim inf" in the relation (6.1) defining the Hausdorff measure can be taken (and therefore the Hausdorff measure computed) by letting $\mathcal{A}$ vary among the almost coverings of A rather than in $\mathcal{C}(A)$.

The next lemma permits us to choose for every set A with finite Hausdorff measure a covering made of disjoint sets and with a sum of diameters proportional to the measure of the set.

Lemma 7.8. *Given any subset A of E such that $\mathcal{H}^{\alpha}(A) < +\infty$ and any $\varepsilon > 0$, one can determine $\bar{\rho} > 0$ such that for every subset B of A and every $\mathcal{B} \in \mathcal{C}(B)$ with $\rho(\mathcal{B}) \leq \bar{\rho}$, one can extract a subset $\mathcal{B}' \subset \mathcal{B}$ consisting of pairwise disjoint sets satisfying*

$$|\mathcal{B}'|_{\alpha} \geq \frac{\mathcal{H}^{\alpha}(B) - \varepsilon}{c_{\alpha} 5^{\alpha}} .$$

Proof. We extract $\mathcal{B}'$ from $\mathcal{B}$ as in Lemma 7.1 and then we see that from Lemma 7.4,

$$(7.9) \qquad\qquad c_{\alpha} |(\mathcal{B}')^{*}|_{\alpha} \geq \mathcal{H}^{\alpha}(B) - \varepsilon$$

if $\rho((\mathcal{B}')^{*})$ is small enough. Since for every B in $\mathcal{B}'$,

$$\text{diam } B^{*} \leq 5 \text{ diam } B ,$$

then $\rho((\mathcal{B}')^{*}) \leq 5 \bar{\rho}$, so (7.9) holds provided $\bar{\rho}$ has been fixed sufficiently small. Moreover,

$$|(\mathcal{B}')^{*}|_{\alpha} \leq 5^{\alpha} |\mathcal{B}'|_{\alpha} ,$$

so from (7.9) we get the thesis. $\square$

By an obvious variant of the preceding proof, we obtain the following corollary.

Corollary 7.10. *If $\mathcal{H}^\alpha(A) = +\infty$, then for any positive constant M we can determine $\overline{p}$ in such a way that every covering $\mathcal{B}$ of A with $\rho(\mathcal{B}) \leq \overline{p}$ contains a disjoint subcovering $\mathcal{B}'$ such that*

$$(7.11) \qquad\qquad |\mathcal{B}'|_\alpha \geq M .$$

Corollary 7.12. *If $\mathcal{H}^\alpha(A) < +\infty$ and c is a positive constant, one can find a constant $\overline{p} > 0$ such that, if $\mathcal{B}$ is any set of subsets of E which covers a part B of A with $\mathcal{H}^\alpha(B) \geq c$ and such that $\rho(\mathcal{B}) \leq \overline{p}$, one can extract a finite subset $\mathcal{B}' \subset \mathcal{B}$ consisting of pairwise disjoint sets such that $|\mathcal{B}'|_\alpha \geq 5^{-(\alpha+1)}c$.*

Proof. It is enough to take $5\varepsilon < 4c$ in Lemma 7.8. $\square$

2. Vitali coverings.

We say that a covering $\mathcal{B}$ of a set A is a *Vitali covering* if, for every x in A, $\mathcal{B}$ contains a set which contains x and has an arbitrary small diameter. Note that if $\mathcal{B}$ is a Vitali covering of A and N is a neighbourhood of A, then the set $\mathcal{B}_N$, defined as the set of the sets in $\mathcal{B}$ which are contained in N, is still a Vitali covering of A. Note also that a covering $\mathcal{B}$ is a Vitali covering if and only if it contains a (Vitali) subcovering $\mathcal{B}_{\overline{p}}$ with $\rho(\mathcal{B}_{\overline{p}}) < \overline{p}$, whatever the choice of $\overline{p} > 0$ is. So the previous result applies to a Vitali covering $\mathcal{B}$ without any restriction on $\rho(\mathcal{B})$.

Vitali coverings are a powerful tool to prove most properties of Hausdorff measurable sets. Indeed, a Vitali covering of a set A contains for every point x of A "windows" of arbitrarily small size which permit to look at what happens to A around x. The Vitali Covering Lemma, to be proved below, will in addition show that Vitali coverings can be made of disjoint sets. Thus, many equalities and inequalities concerning the Hausdorff measure of a given set A will be obtainable by simply proving them *locally*, on sets of a Vitali covering of A and thereafter summing them.

We say that a set of sets $\mathcal{B}$ is a *Vitali approximate covering* of A if one can determine a Vitali covering of A, $\hat{\mathcal{B}}$, and a positive number $\varepsilon > 0$ such that

$$(7.13) \qquad \begin{array}{l} \forall\, \hat{B} \in \hat{\mathcal{B}} \;\; \exists\, B \in \mathcal{B} \quad \text{such that} \quad B \subset \hat{B} \quad \text{and} \\[2mm] \operatorname{diam}(B) \geq \varepsilon \operatorname{diam}(\hat{B}) . \end{array}$$

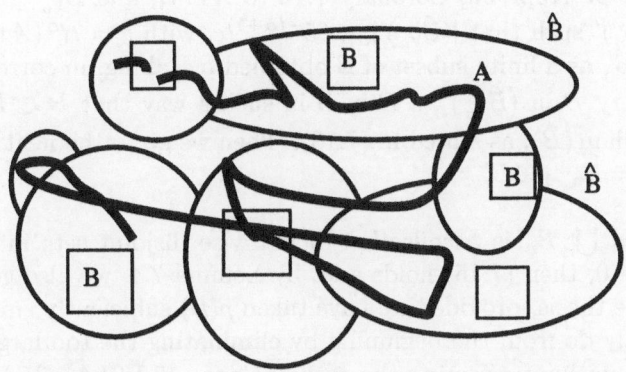

Figure 7.1 : in bold, a set A. The ellipses are elements
of a Vitali covering of A, while the squares are the
corresponding elements of a Vitali approximate covering.

Vitali approximate coverings are a very useful generalization of Vitali coverings. Indeed, many properties of Vitali coverings remain true. Of course, an approximate Vitali covering of A is not a covering of A (see Figure 7.1). However, approximate coverings can be used to give an account of the Hausdorff measure of the set which they approximately "cover". A (main) use of Vitali approximate coverings is to be stated now.

Lemma 7.14 (Vitali Covering Lemma). *Let* $\mathcal{H}^{\alpha}(A) < +\infty$, $\mathcal{B}$ *be an approximate Vitali covering of* A *made of closed sets and* $\bar{\rho} > 0$. *Then, for every number* $b < \mathcal{H}^{\alpha}(A)$, $\mathcal{B}$ *contains a countable subset* $\mathcal{A}$ *consisting of pairwise disjoint sets such that* $\rho(\mathcal{A}) < \bar{\rho}$ *and*

$$c_{\alpha}|\mathcal{A}|_{\alpha} > b,$$

with the following alternative: either $|\mathcal{A}|_{\alpha} = +\infty$ *or*

$$(7.15) \qquad \mathcal{H}^{\alpha}(A \setminus \bigcup \mathcal{A}) = 0.$$

Proof. Let $\hat{\mathcal{B}}$ be a Vitali covering of A such that (7.13) holds. We define recursively an increasing sequence of finite subsets $\mathcal{B}_n$ of $\mathcal{B}$ in the following way.

We set $\mathcal{B}_0 = \emptyset$ and, if $\mathcal{B}_{n-1}$ is given, we consider the open set V_n defined by $V_n = E \setminus \bigcup \mathcal{B}_{n-1}$. As noticed at the beginning of this section,

$\hat{B}$ contains a part $\hat{B}_{V_n}$ which still is a Vitali covering of $A \cap V_n$ and satisfies $\rho(\hat{B}_{V_n}) < \bar{\rho}$. Applying Corollary 7.12 to $A \cap V_n$ and $\hat{B}_{V_n}$, we find a covering $(\hat{B}_{V_n})'$ such that $|(\hat{B}_{V_n})'|_\alpha > 5^{-(\alpha+1)}c$, with $c = \mathcal{H}^\alpha(A \cap V_n)$. We construct B'_n as a finite subset of B obtained by taking, in correspondence with every $\hat{B}$ in $(\hat{B}_{V_n})'$, a B in B in such a way that $B \subset \hat{B}$ and diam $(B) \geq \varepsilon$ diam $(\hat{B})$, as stated in (7.13). Then we go to the next index by setting $B_n = B_{n-1} \cup B'_n$.

Clearly $\mathcal{A} = \bigcup_n B_n$ is a collection of pairwise disjoint sets in B. If $\mathcal{H}^\alpha(A \cap V_n) \to 0$, then (7.15) holds and, by Lemma 7.4, we also get the first part of the thesis, provided we have taken $\rho(B)$ sufficiently small, as we can certainly do from the beginning by eliminating the too large sets from B and $\hat{B}$ without affecting the assumptions. If $\mathcal{H}^\alpha(A \cap V_n) \not\to 0$, since, by construction,

$$|B'_n|_\alpha \geq \varepsilon^\alpha |(\hat{B}_{V_n})'|_\alpha \geq \varepsilon^\alpha 5^{-(\alpha+1)} \mathcal{H}^\alpha(A \cap V_n) \,,$$

we have

$$|\mathcal{A}|_\alpha = \sum_n |B'_n|_\alpha = +\infty \,. \quad \square$$

3. Hausdorff and Lebesgue measure.

As an example of the use of the lemmas in this section we can easily prove that if $E = \mathbb{R}^N$ and $\alpha = N$, then $\mathcal{H}^\alpha$ agrees with the Lebesgue outer measure $\mathcal{L}$ of dimension N, as claimed at the end of the last chapter. Recall that the Lebesgue outer measure is defined by first fixing the measure of a N-dimensional cube C_λ with edge length λ to be equal to λ^N. Then the Lebesgue outer measure $\mathcal{L}(A)$ of a set $A \subset \mathbb{R}^N$ is defined as the infimum of the numbers $\sum_i \lambda_i^N$, taken over all families of cubes $(C_{\lambda_i})_i$ covering A. It is easily seen that the abstract theory of outer and Carathéodory measures developed in Chapter 6 applies to $\mathcal{L}$. In order to prove that Lebesgue and Hausdorff measure agree, we only need to admit the so called *isodiametric inequality* [Egg]. It states that every set in $\mathbb{R}^N$ has a Lebesgue measure less than or equal to the measure of the ball with the same diameter. In other words

(7.16) $\forall\, X \subset \mathbb{R}^N :\quad \mathcal{L}(X) \leq c_N (\text{diam } X)^N \,.$

(Recall that we chose the normalisation constant c_N in such a way that N-dimensional Lebesgue measure and Hausdorff measure agree on balls.)

So we have for every $A \subset \mathbb{R}^N$ and $\mathcal{A} \in \mathcal{C}(A)$:

$$(7.17) \qquad\qquad c_N \, |\mathcal{A}|_N \ \geq \ \sum_{X \in \mathcal{A}} \mathcal{L}(X) \ \geq \ \mathcal{L}(A)$$

because of the countable subadditivity of $\mathcal{L}$. We can pass to the limit in (7.17) and we get

$$(7.18) \qquad\qquad \forall \, A \subset \mathbb{R}^N : \quad \mathcal{L}(A) \ \leq \ \mathcal{H}^N(A) \, .$$

On the other hand, if A is any open set of $\mathbb{R}^N$, we can apply the Vitali Covering Lemma 7.14 in the following way.

For every positive constant $b < \mathcal{H}^N(A)$, we can determine a set $\mathcal{A}$ of disjoint balls contained in A such that $c_N |\mathcal{A}|_N \geq b$. (It is enough to apply the lemma starting with the set $\mathcal{B}$ of all the closed balls contained in A.) Then

$$(7.19) \quad b \leq c_N |\mathcal{A}|_N \ = \ \sum_{B \in \mathcal{A}} c_N \, (\mathrm{diam}\, B)^N \ = \ \sum_{B \in \mathcal{A}} \mathcal{L}(B) \ \leq \ \mathcal{L}(A) \, .$$

By (7.18) and (7.19) we see that the two outer measures coincide.

Chapter 8

DENSITY PROPERTIES

In this chapter and the following, we shall be concerned with properties which are true for all points of a given set A, except a set of points with zero Hausdorff measure. In this situation, we say that the considered property is true almost everywhere (a.e.) in K. We call α-set any $\mathcal{H}^\alpha$-measurable subset of a metric space E such that $\mathcal{H}^\alpha(K) < +\infty$. We shall begin to develop the Besicovitch theory of sets with finite Hausdorff measure.

In order to explain the relevance of the results proved below, let us take as an (important) example, the case $\alpha = 1$. Then obvious examples of 1-sets are the differentiable Jordan curves with finite length. It is easily seen that for such sets A, the length contained in a ball $B(x, r)$ centered at a point x of A is equivalent to 2r as r tends to zero. This relation is true for all points of A except for the endpoints of the curve. Let us take an example of quite different structure: the "planar Cantor set" defined in Section 4 of Chapter 4. Let us denote this set by K. It is easily seen (and we prove at the end of the present chapter) that the "spherical lower density" $\liminf_{r \to 0} \frac{\mathcal{H}^1(K \cap B(x,r))}{2r}$ is not 1. So we observe a quite different spatial repartition for K and for smooth curves. It is expected that a general 1-set will have points where "curve-like" behaviour occurs (the spherical density is 1) and points where "Cantor-like" behaviour occurs. The first kind of point will be called "regular" and the other one "irregular". One of the most remarkable results of the Besicovitch theory is that this classication is stable when one cuts a set into measurable pieces: The set of the regular points of an α-set K which are contained in a measurable subset J is essentially the same as the set of regular points of J!

The same thing happens for irregular points. So the properties of "regularity" and "irregularity" are intrinsic, local, stable properties. Every set will be divided by the Besicovitch theory into its regular part (which we will later prove to be contained in a bunch of rectifiable curves) and its irregular part (which will be, under many aspects, similar to the planar Cantor set). Those parts clearly behave like two nonmiscible fluids, and the Besicovitch theory yields clear computational criteria (the local spherical densities) to decide whether a point is regular or irregular. From the Image Processing viewpoint, we can say that the Besicovitch theory yields an absolute definition of which points in an "edge map" are really points of an edge and which are not (the irregular points).

1. Upper and lower density.

We start the theory with general density results. Let $A, K \subset E$. We shall denote by $\mathcal{D}_K^\alpha(A)$ (or, when there is no ambiguity about K, $\mathcal{D}^\alpha(A)$) the $\mathcal{H}^\alpha$-mean density of K on A defined as

$$(8.1) \qquad \mathcal{D}_K^\alpha(A) = \frac{\mathcal{H}^\alpha(K \cap A)}{c_\alpha \, (\mathrm{diam} \, A)^\alpha} \, .$$

Lemma 8.2. *Let K be a subset of E such that $\mathcal{H}^\alpha(K) < +\infty$. Then for a.e. x in K,*

$$(8.3) \qquad \limsup_{\substack{A \ni x \\ \mathrm{diam}\,(A) \to 0}} \mathcal{D}_K^\alpha(A) = 1 \, .$$

Proof. We shall begin the proof by showing that the left-hand side of the above inequality is less than or equal to 1. As we shall now prove, this is an easy consequence of Vitali Lemma 7.14 while the other inequality directly derives from the definition of Hausdorff measure. Let K_n be the set of the points x in K such that the left-hand side of (8.3) is larger than $1 + \frac{1}{n}$. By this definition we can find a Vitali covering $\mathcal{B}$ of K_n such that

$$(8.4) \qquad \forall \, B \in \mathcal{B} \, : \quad \mathcal{D}_K^\alpha(B) \geq 1 + \frac{1}{n} \, .$$

By Vitali Lemma 7.14, there is a subset $\mathcal{B}'$ of $\mathcal{B}$ that consists of disjoint sets such that

$$(8.5) \qquad c_\alpha |\mathcal{B}'|_\alpha \geq \frac{1}{2} \mathcal{H}^\alpha(K_n) \, .$$

(We take the constant $\frac{1}{2}$ in the above estimate for simplicity of notation; we could take any fixed positive constant smaller than 1.) By adding the terms in (8.4) obtained for every B in $\mathcal{B}'$ and using (8.1), we get

$$(8.6) \qquad \mathcal{H}^\alpha\!\left(K \cap \bigcup \mathcal{B}'\right) \geq \left(1 + \frac{1}{n}\right) c_\alpha |\mathcal{B}'|_\alpha \, .$$

On the other hand, by the approximation Lemma 7.4, for any given $\varepsilon > 0$ we can take $\mathcal{B}$ with $\rho(\mathcal{B})$ small enough to ensure

$$(8.7) \qquad \mathcal{H}^\alpha\!\left(K \cap \bigcup \mathcal{B}'\right) \leq c_\alpha |\mathcal{B}'|_\alpha + \varepsilon \, .$$

Therefore by (8.5) and by (8.6)-(8.7) we get

$$\mathcal{H}^\alpha(K_n) \leq 2c_\alpha |\mathcal{B}'|_\alpha \leq 2n\varepsilon \,.$$

By the arbitrariness of ε we have $\mathcal{H}^\alpha(K_n) = 0$ and so by the subadditivity of the Hausdorff measure (6.3) one sees that $\mathcal{H}^\alpha(K') = 0$ where K' stands for the set of the points x of K such that the strict inequality ">" holds instead of the equality in (8.3).

Let now $K_{\infty,n}$ be the set of the points x in K such that the left-hand side of (8.3) is strictly bounded from above by $\left(1 - \frac{1}{n}\right)$. Then we can write

$$(8.8) \qquad\qquad K_{\infty,n} = \bigcup_m K_{m,n} \,,$$

where the sets $K_{m,n}$ are taken in such a way that

$$(8.9) \qquad \forall\, x \in K_{m,n} \,, \ \forall\, A \subset E \,, \ \text{if } x \in A \text{ and } \operatorname{diam} A \leq \frac{1}{m} \ :$$
$$\mathcal{D}_K^\alpha(A) \leq 1 - \frac{1}{n} \,.$$

By the definition of Hausdorff measure (6.1), we can find coverings $\mathcal{A}_i$ in $\mathcal{C}(K_{m,n})$ such that

$$\rho(\mathcal{A}_i) \to 0 \quad \text{and} \quad c_\alpha |\mathcal{A}_i|_\alpha \to \mathcal{H}^\alpha(K_{m,n}) \,.$$

We can assume that $\rho(\mathcal{A}_i) < \frac{1}{m}$ and so by (8.9)

$$(8.10) \qquad \forall\, i \ : \ \mathcal{H}^\alpha(K_{m,n}) \leq \mathcal{H}^\alpha(K \cap \bigcup \mathcal{A}_i) \leq \left(1 - \frac{1}{n}\right) c_\alpha |\mathcal{A}_i|_\alpha$$
$$\to \left(1 - \frac{1}{n}\right) \mathcal{H}^\alpha(K_{m,n}) \,.$$

So, for every choice of m, n, $\mathcal{H}^\alpha(K_{m,n}) = 0$ and therefore, by the subadditivity of the Hausdorff measure (6.3), $\mathcal{H}^\alpha(K_{\infty,n}) = 0$ and, subsequently, $\mathcal{H}^\alpha(K'') = 0$, where $K'' = \bigcup_n K_{\infty,n}$ is the set of the points x in K such that the strict inequality $<$ holds in (8.3). $\quad\square$

The next lemma states that every α-set A somehow is "essentially closed". In fact, we are going to see that, with the exception of a set with zero α-dimensional measure, the upper density of the set at the points which are outside A is 0. Therefore almost all such points are "essentially" out of the closure of A. Indeed, there are balls centered in them where A has an arbitrarily small density. This gives a strong computational method to decide whether or not a point belongs to an α-set: by Lemma 8.2, the upper density inside A is 1, while by the next lemma it is zero outside!

Lemma 8.11. *Let K be an α-set. Then for a.e. x in $E \setminus K$,*

$$(8.12) \qquad \limsup_{\substack{x \in A \\ \text{diam}\,(A) \to 0}} \mathcal{D}_K(A) \, = \, 0 \,.$$

Proof. Let H_n be the set of the points x in $E \setminus K$ such that the left-hand side of (8.12) is stricly larger than $\frac{1}{n}$ and let C be any closed subset of K. By definition, H_n has a Vitali covering $\mathcal{B}$ such that

$$(8.13) \qquad \forall\, B \in \mathcal{B} \, : \;\; \mathcal{D}_K(B) \, \geq \, \frac{1}{n} \,,$$

By the Vitali Covering Lemma 7.14 we can find a subset $\mathcal{B}'$ of $\mathcal{B}$ consisting of disjoint sets, contained in the open neighbourhood of H_n, $E \setminus C$, and such that

$$(8.14) \qquad c_\alpha |\mathcal{B}'|_\alpha \, \geq \, \frac{1}{2} \mathcal{H}^\alpha(H_n) \,.$$

By (8.13) we have

$$(8.15) \qquad \mathcal{H}^\alpha(K \setminus C) \, \geq \, \mathcal{H}^\alpha\Big(K \cap \bigcup \mathcal{B}'\Big) \, \geq \, \frac{1}{n} c_\alpha |\mathcal{B}'|_\alpha \,.$$

From (8.14)-(8.15) we obtain

$$(8.16) \qquad \mathcal{H}^\alpha(H_n) \, \leq \, 2n\, \mathcal{H}^\alpha(K \setminus C) \,.$$

By Proposition (6.26), the measure of an α-set is the supremum of the measures of the closed sets which it contains. So we can let $\mathcal{H}^\alpha(K \setminus C) \to 0$ and we get, from (8.16), $\mathcal{H}^\alpha(H_n) = 0$ and therefore $\mathcal{H}^\alpha(H') = 0$, where $H' = \bigcup_n H_n$ is the set of the points x in $E \setminus K$ such that the left-hand side of (8.12) is positive. $\quad\square$

2. Regular sets.

We can now pass to the study of a more geometrical concept: the spherical density. Let K be a subset of E. For every x in E we define the lower spherical density of K at a point x, $\underline{d}_K^\alpha(x)$, and the upper spherical density of K, $\overline{d}_K^\alpha(x)$, by

$$\underline{d}_K^\alpha(x) = \liminf_{R\to 0} \mathcal{D}_K(B_R(x))$$

$$\overline{d}_K^\alpha(x) = \limsup_{R\to 0} \mathcal{D}_K(B_R(x)).$$

Corollary 8.17. *If $\mathcal{H}^\alpha(K) < +\infty$, then for every x in E,*

$$0 \leq \underline{d}_K^\alpha(x) \leq \overline{d}_K^\alpha(x)$$

and for a.e. x in K,

(8.18) $2^{-\alpha} \leq \overline{d}_K^\alpha(x) \leq 1$.

If, in addition, K is $\mathcal{H}^\alpha$-measurable, then for a.e. x in $E \setminus K$,

(8.19) $\underline{d}_K^\alpha(x) = \overline{d}_K^\alpha(x) = 0$.

Proof. Of course, (8.19) and (8.18) follow from Lemmas (8.11) and (8.2) respectively, since every set A which contains x is contained in the ball centered at x with radius diam(A).   □

When $\underline{d}_K^\alpha(x)$ and $\overline{d}_K^\alpha(x)$ are equal we shall call $\mathcal{H}^\alpha$-*density* of K at x their common value and we shall denote it by $d_K^\alpha(x)$. The following two corollaries are obvious consequences of (8.19).

Corollary 8.20. *Let A, B be two α-sets of E such that $A \subset B$. Then for a.e. x in A*

(8.21) $\underline{d}_A^\alpha(x) = \underline{d}_B^\alpha(x)$ $\overline{d}_A^\alpha(x) = \overline{d}_B^\alpha(x)$.

Proof. We just have to notice that $d_{B\setminus A}^\alpha = 0$ for a.e. x in A and that by the additivity of the Hausdorff measure on A and B,

$$\underline{d}_B^\alpha(x) = \underline{d}_A^\alpha(x) + d_{B\setminus A}^\alpha(x), \qquad \overline{d}_B^\alpha(x) = \overline{d}_A^\alpha(x) + d_{B\setminus A}^\alpha(x).$$

Corollary 8.22. *Let A, B be two α-sets. Then (8.21) holds for a.e. x in $A \cap B$.*

Proof. In fact, both $\underline{d}_A^\alpha(x)$ and $\underline{d}_B^\alpha(x)$ are equal, by the previous corollary, to $\underline{d}_{A \cap B}^\alpha(x)$. $\quad\square$

A set A is called *regular* if

$$(8.23) \qquad\qquad \text{for a.e. } x \text{ in } A\text{: } \underline{d}_A^\alpha(x) = 1 .$$

By (8.18) the above equality is equivalent to the inequality $\underline{d}_A^\alpha(x) \geq 1$ for a.e. x in A if $\mathcal{H}^\alpha(A) < +\infty$. On the other hand, we shall call A *fully irregular* if the equality in (8.23) is not satisfied for a.e. x in A or, equivalently, if A does not contain any regular subset with nonzero outer measure.

From the previous corollaries we easily obtain

Corollary 8.24. *Every measurable subset of a regular measurable subset A of E is regular. Every subset of a fully irregular subset A of E is fully irregular.*

Corollary 8.25. *If A and B are two α-sets, if A is regular and B is fully irregular, then $A \cap B$ is a negligible subset of E.*

For $A \subset \mathbb{R}^N$ and $n, m \in \mathbb{N}$, we shall denote in the following by $A_{n,m}$ the set of the points x in $\mathbb{R}^N$ such that for every $r < \frac{1}{m}$ we have

$$(8.26) \qquad\qquad 1 - \frac{1}{n} \leq \mathcal{D}_A(B_r(x)) \leq 1 + \frac{1}{n}$$

and for every $X \subset \mathbb{R}^N$, $x \in X$, $\operatorname{diam}(X) < \frac{1}{m}$

$$(8.27) \qquad\qquad \mathcal{D}_A(X) \leq 1 + \frac{1}{n} .$$

Proposition 8.28. *A set A is regular if and only if for every value of n, $(A_{n,m})_{m \in \mathbb{N}}$ is an almost covering of A.*

Proof. This follows from the definition of regularity since, for every regular point x in A and whatever n is, (8.26) and (8.27) must hold if m is sufficiently small. $\quad\square$

How to prove that sets defined from density properties are measurable.

We shall now prove that every $A_{n,m}$ is a Borel set and is therefore $\mathcal{H}^\alpha$-measurable. Our aim also is to show how measurability arguments work. After this only exception, we shall never give the proof of the measurability of the sets which we shall define in this book. The reason is that such proofs usually require a technical disscussion. We wish to avoid the risk of interrupting the flow of the main arguments for checking details which always use the same ideas in the same way. From now on, analogous proofs will therefore always be left as an exercise to the interested reader.

Proposition 8.29. *The sets $A_{m,n}$ are Borel sets and therefore Hausdorff measurable for every α.*

Proof. First we observe that, in order to show the measurability of the sets $A_{m,n}$, we can assume that the inequalities in (8.26)-(8.27) are replaced by the corresponding strict cases. In fact, the set of the points x such that a certain condition $f(x) \geq a$ is satisfied is the countable intersection of the sets of the points x such that $f(x) > a - \frac{1}{n}$, where $n \in \mathbb{N}$. If all such sets are measurable, then so is the set $\{x, f(x) \geq a\}$.

Conversely, the set of points x where $f(x) > a$ can be thought of as the countable union of the sets $\{x, f(x) \geq a + \frac{1}{n}\}$. Hence also the strict inequalities ">" in the definition of the sets can be replaced by inequalities "$\geq$" with the aim of checking the measurability.

We begin by proving that the set M of the points x which satisfy (8.27) for every $X \subset \mathbb{R}^N$ containing x and such that $\mathrm{diam}(X) < \frac{1}{m}$ is measurable. Indeed, in the definition of M we can impose that X is open, because if $\mathrm{diam}(X) < \frac{1}{m}$, then for $\varepsilon < \frac{1}{2}(\frac{1}{m} - \mathrm{diam}(X))$ the open set $X^\varepsilon = \{x, d(x, X) < \varepsilon\}$ also satisfies $\mathrm{diam}(X^\varepsilon) < \frac{1}{m}$. Thus the set $E \setminus M$ turns out to be equal to the union of all the open sets that have a diameter smaller than $\frac{1}{m}$ and such that (8.27) does not hold. It is therefore open, and M is closed and measurable. In the same way, fix a point x such that the first inequality in (8.26) does not hold for some value of $r \leq \frac{1}{m}$, that is

$$\mathcal{H}^\alpha(A \cap B_r(x)) < 2(1 - \frac{1}{n})r.$$

Then for some $r' < r$ we still have $\mathcal{H}^\alpha(A \cap B_r(x)) < 2(1 - \frac{1}{n})r'$. Since for all points $y \in B_{r-r'}$ we have $B_{r'}(y) \subset B_r(x)$, we deduce that for such points y,

$$\mathcal{H}^\alpha(A \cap B_{r'}(x)) < 2(1 - \frac{1}{n})r'.$$

So the set of the points which do not satisfy the inequality is open. Analogously, if for x and r the second inequality in (8.26) does not hold, we can fix $r' > r$ such that r' still is smaller than $\frac{1}{m}$ and $\mathcal{H}^1(A \cap B_r(x)) \geq 2(1 + \frac{1}{n}r')$. Also this inequality is therefore violated, in correspondence with r' for the points in $B_{r-r'}(x)$, and we can conclude as before. Note that this inequality would be implicit, by (8.27), for $r < \frac{1}{2m}$. $\square$

An example of a fully irregular 1-set.

We shall end this chapter with an example of irregular set. Regular sets are at hand since for instance every k-dimensional affine subspace of $\mathbb{R}^N$ is regular. Irregular sets require a construction. We shall prove that the "planar Cantor set" B defined at the end of Chapter 6 is fully irregular.

Proposition 8.30. *The planar Cantor set B defined in Section 6.4 has everywhere a lower spherical density less than $\frac{1}{\sqrt{2}}$.*

Proof. Indeed, if we consider the three triangles in $\mathcal{T}_1$ (with the notation of section 6.4) represented in Figure 6.2, we can easily check that the distance between two of them is always larger or equal to $\frac{\sqrt{2}}{3}$. So if we consider any disk D with radius $\frac{\sqrt{2}}{3}$, centered at a point of B, and therefore at a point belonging to one of the triangles in $\mathcal{T}_1$, it can intersect only one of such triangles. This fact shows that $\mathcal{H}^1(D \cap B) \leq \frac{1}{3}$ and therefore that the mean density $\mathcal{D}^1_B(D)$ is less or equal to $\frac{1}{\sqrt{2}}$. One can easily repeat this argument at arbitrarily small scales, by observing that the minimum distance between two distinct triangles in $\mathcal{T}_n$ is larger or equal to $\frac{\sqrt{2}}{3^n}$. So, if we take as D any ball centered at a point of B with radius $\frac{\sqrt{2}}{3^n}$, the same estimate as above holds. This fact shows that the lower spherical density of K at each one of its points is bounded from above by $\frac{1}{\sqrt{2}}$. $\square$

Chapter 9

TANGENCY PROPERTIES OF
REGULAR SUBSETS OF $\mathbb{R}^N$

In this chapter, we shall be concerned with some tangency properties of regular α-sets of $\mathbb{R}^N$. We shall prove that α is necessarily an integer and that a regular set admits a tangent affine space almost everywhere and therefore "looks like" an α-dimensional manifold almost everywhere. The proof is essentially based on a remarkable reflection lemma due to Marstrand. Roughly speaking, this lemma, which is not difficult to prove once the right ideas are at hand, says that given two points x and y in a regular set A, the reflected point $2x - y$ lies close to A. This argument can be iterated by successive reflections and therefore yields nets of points which are contained in an affine subspace V and close to A. Of course, the whole argument is a little bit more technical than this abstract. We shall need precise estimates on how a net of points obtained by reflection spreads out and on how close to A the reflected point $2x - y$ is. This will be done in the first section. The second section is devoted to the existence proof, for any $r > 0$, of "approximate tangent affine spaces", $V(x,r)$, at most points $x \in A$. The third section analyses accurately how close A is to its approximate tangent spaces. It is proved that the orthogonal projection of A onto $V(x,r)$ has a density inside $B(x,r)$ tending to 1. This fact easily implies convergence of the sets $V(x,r)$ towards an (almost everywhere) unique tangent space $V(x)$.

1. Reflection lemmas.

If A is a subset of $\mathbb{R}^N$ we shall denote by $A^{(1)}$ the *reflection of A*, defined as the set of all points $2x - y$ such that x and y are in A. Then we set by recursion

$$A^{(k+1)} = (A^{(k)})^{(1)}$$

and in this way we get a monotone sequence, by inclusion, of subsets of $\mathbb{R}^N$. Of course, for every h in $\mathbb{N}$, $A^{(h)}$ is contained in the affine subspace of $\mathbb{R}^N$ spanned by A, denoted by $\mathcal{S}(A)$. We can easily observe that

$$(9.1) \qquad\qquad \operatorname{diam}(A^{(1)}) \leq 3 \operatorname{diam}(A).$$

For n in $\mathbb{N}$, let $A_{(n)}$ be the set of the points $mx - (m-1)y$ obtained in correspondence with an integer $m \leq n+1$ and with two points x and y in A. Since

$$mx - (m-1)y = 2((m-1)x - (m-2)y) - ((m-2)x - (m-3)y),$$

one sees that $A_{(n+1)} \subset (A_{(n)})^{(1)}$ and therefore

(9.2) $\forall n \in \mathbb{N} : \quad A_{(n)} \subset A^{(n)} .$

The two sets in (9.2) are equal when $n = 1$ and are, in general, different for $n > 1$. Our interest in distinguishing $A_{(n)}$ as a subset of $A^{(n)}$ in the next lemmas is that we can compute a bound on $\mathrm{diam}\,(A_{(n)})$ which is much better than the bound on $\mathrm{diam}\,(A^{(n)})$ that comes out from the iteration of (9.1). Indeed, one obviously has

$$\forall n \in \mathbb{N} : \quad \mathrm{diam}\,(A_{(n)}) \leq (2n+1)\,\mathrm{diam}\,(A) .$$

The above inequality, combined with a suitable iteration of (9.1), gives

(9.3) $\forall n, k \in \mathbb{N} : \quad \mathrm{diam}\,((A_{(n)})^{(k)}) \leq 3^k (2n+1)\,\mathrm{diam}\,(A) .$

The estimate given in the next lemma is not sharp but tries to establish a simple lower bound to the number of points of $(A_{(n)})^{(k)}$. By $d(x, V)$, we denote the *euclidean distance from a point x to an affine subspace V*.

Lemma 9.4 (First Geometric Lemma). *Let k be an integer and δ a real positive number. Let $x_0, x_1, \ldots, x_k$ be a finite sequence of points of $A \subset \mathbb{R}^N$ such that*

(9.5) $\forall h = 1, 2, \ldots, k : \quad d(x_h, S_h) \geq \delta ,$

where S_h stands for the affine space $S(\{x_0, x_1, \ldots x_{h-1}\})$ spanned by $x_0, x_1, \ldots, x_{h-1}$. Then, for any integer n, $(A_{(n)})^{(k)}$ contains at least n^k points having a distance larger or equal to δ from each other.

Proof. If $k = 0$, the statement is obvious and we can assume by induction that the theorem is true if k is replaced by $k - 1$.

So $(\{x_0, x_1, \ldots, x_{k-1}\}_{(n)})^{(k-1)}$ contains n^{k-1} points $y_1, y_2, \ldots, y_{n^{k-1}}$ as required in the statement. Moreover, $\{x_0, x_k\}_{(n)}$ contains the points

$$z_1 = x_k , \; z_2 = 2x_k - x_0 , \; z_3 = 3x_k - 2x_0 , \ldots, \; z_n = nx_k - (n-1)x_0 .$$

The n sets $2S_k - z_i$ $(i = 1, 2, \ldots, n)$ are n parallel affine subspaces having a distance larger or equal to δ from each other. Indeed, denoting by V the linear space generated by $x_1 - x_0 , \ldots, x_{k-1} - x_0$, we have $S_k = x_0 + V$ and $d(x_0 - x_k, V) \geq \delta$. Thus, since $2V = V$, $d(2S_k - z_1, 2S_k - z_2) = d(2x_0 - z_1 + V, 2x_0 - z_2 + V) = d(2x_0 - x_k + V, 3x_0 - 2x_k + V) = d(V, x_0 - x_k + V) = d(V, x_0 - x_k) \geq \delta$ (by the assumption (9.5)). So the parallel spaces $2S_k - z_i$ are at a distance at least δ from each other and $2S_k - z_i$ contains the n^{k-1} points $2y_j - z_i$, $j = 1, 2, \ldots, n^{k-1}$, which have by assumption a distance larger or equal to 2δ from each other. Since all those points are by construction in $(A_{(n)})^{(k)}$, the theorem is proved. $\square$

Let x and y be two points in $\mathbb{R}^N$ and d their distance. Let ε be a real number, $0 < \varepsilon < 1$, and let B be the ball centered at x with radius $(1 - \varepsilon)d$ and B' the ball centered at y with radius εd. The balls B and B' are clearly tangent. Let y' be the point of B opposite to the tangency point with B', and let B'' be the ball centered at y' with radius $3\sqrt{\varepsilon}d$.

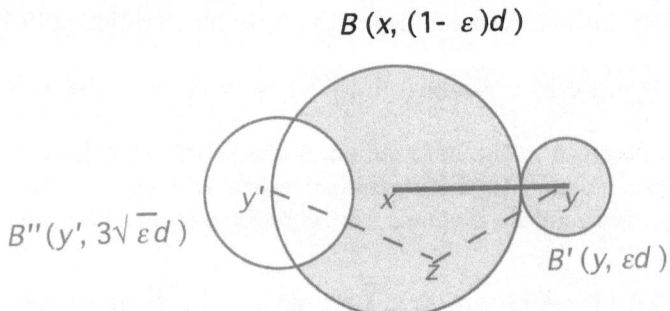

$$B\,(x,\,(1-\,\varepsilon)d\,)$$

$B''(y',\,3\sqrt{\varepsilon}\,d\,)$

$B'\,(y,\,\varepsilon d\,)$

Figure 9.1 : Second Geometric Lemma

Lemma 9.6 (Second Geometric Lemma). *Let $B, B', B'', \varepsilon, d$ be as defined above. Then*

$$\operatorname{diam}\left((B \cup B') \setminus B''\right) \;\leq\; 2(1 - \varepsilon)d = \operatorname{diam}(B).$$

Proof. In fact (see Figure 9.1), if we take into account that $y' - z$ has a positive scalar product with $z - y$ for every z in B, we have

$$\forall\, z \in B \setminus B'' \qquad d(y, y')^2 \;\geq\; d(y, z)^2 + d(y', z)^2$$
$$\geq\; d(y, z)^2 + (3\sqrt{\varepsilon}d)^2.$$

Thus

$$d(y, z)^2 \;\leq\; (2 - \varepsilon)^2 d^2 - 9\varepsilon d^2$$
$$\leq\; (2 - 3\varepsilon)^2 d^2 \,.$$

Therefore, for any z in $B \setminus B''$ and z' in B', $d(z, z') \leq d(z, y) + d(y, z') \leq (2 - 3\varepsilon)d + \varepsilon d$. $\square$

In the following, we consider an α-set $A \subset \mathbb{R}^N$ and, for $n, m \in \mathbb{N}$, we denote by $A_{n,m}$ the set of the points x in $\mathbb{R}^N$ such that for every $r < \frac{1}{m}$ we have

$$(9.7) \qquad\qquad 1 - \frac{1}{n} < \mathcal{D}_A(B_r(x)) < 1 + \frac{1}{n}$$

and for every $X \subset \mathbb{R}^N$, $x \in X$, $\mathrm{diam}\,(X) < \frac{1}{m}$

$$(9.8) \qquad\qquad \mathcal{D}_A(X) < 1 + \frac{1}{n}\,.$$

We have seen at the end of the last chapter (Propositions 8.28 and 8.29) that $A_{n,m}$ is a Borel set and is therefore $\mathcal{H}^\alpha$-measurable. Moreover, A is regular if and only if for every value of n, $(A_{n,m})_{m \in \mathbb{N}}$ is an almost covering of A.

We define the closed δ-neigbourhood of a set A by

$$\mathcal{N}_\delta(A) = \{x, d(x, A) \le \delta\}.$$

Lemma 9.9 (First Reflection Lemma). *Let A be a regular α-set of $\mathbb{R}^N$. Then for every $\delta > 0$ one can find a number $\overline{d} > 0$ and a subset A' of A such that*

$$(9.10) \qquad\qquad \mathcal{H}^\alpha(A \setminus A') < \delta$$

and for every $B \subset A'$ with $\mathrm{diam}\,(B) = d \le \overline{d}$,

$$(9.11) \qquad\qquad B^{(1)} \subset \mathcal{N}_{\delta d}(A)\,.$$

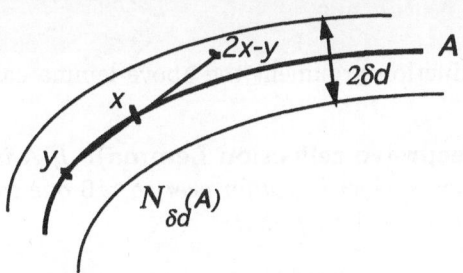

Figure 9.2 : First Reflection Lemma

Proof. We shall take $A' = A_{n,m} \cap A$ for suitable values of n and m, so that (9.10) is ensured for a given n if m is large enough. Assume by contradiction that (9.11) does not hold for a value of $\bar{d}$ less than $\frac{1}{2m}$. Then we can find x and y in A' such that $d(x,y) = d \leq \bar{d}$ and

$$(9.12) \qquad\qquad 2x - y \notin \mathcal{N}_{\delta d}(A) \,.$$

We fix $\varepsilon > 0$ in such a way that $3\sqrt{\varepsilon} < \delta$. Then, if we consider the balls B, B' and B'' defined in Second Geometric Lemma 9.6, (9.12) implies

$$B(2x - y, 3\sqrt{\varepsilon}d) \cap A = B'' \cap A = \emptyset.$$

Thus

$$(9.13) \qquad \mathcal{H}^\alpha(A \cap ((B \cup B') \setminus B'')) \;=\; \mathcal{H}^\alpha(A \cap B) + \mathcal{H}^\alpha(A \cap B') \,.$$

Since x and y are in $A_{n,m}$ we have, using Lemma 9.6 and (9.7), (9.8),

$$\mathcal{H}^\alpha(A \cap ((B \cup B') \setminus B'')) \;\leq\; c_\alpha (1 + \frac{1}{n}) \,\mathrm{diam}\, ((B \cup B') \setminus B'')^\alpha$$

$$\leq\; c_\alpha 2^\alpha \left(1 + \frac{1}{n}\right)(1 - \varepsilon)^\alpha d^\alpha;$$

$$\mathcal{H}^\alpha(A \cap B) \;\geq\; c_\alpha \left(1 - \frac{1}{n}\right)(\mathrm{diam}\, B)^\alpha \;=\; c_\alpha 2^\alpha \left(1 - \frac{1}{n}\right)(1 - \varepsilon)^\alpha d^\alpha;$$

$$\mathcal{H}^\alpha(A \cap B') \;\geq\; c_\alpha \left(1 - \frac{1}{n}\right)(\mathrm{diam}\, B')^\alpha \;=\; c_\alpha 2^\alpha \left(1 - \frac{1}{n}\right)\varepsilon^\alpha d^\alpha \,.$$

Therefore (9.13) yields

$$\left(1 - \frac{1}{n}\right)\left(1 + \left(\frac{\varepsilon}{1-\varepsilon}\right)^\alpha\right) \;\leq\; 1 + \frac{1}{n}$$

which is certainly false if n is conveniently large. $\square$

By a simple induction argument the above lemma can be iterated in the following way.

Lemma 9.14 (Recursive reflection Lemma). *If A is a regular α-set of $\mathbb{R}^N$, then for every integer k and for every $\delta > 0$ one can find a number $\bar{d} > 0$ and a subset A' of A such that*

$$\mathcal{H}^\alpha(A \setminus A') \;<\; \delta$$

and that for every $B \subset A'$ with $\mathrm{diam}\,(B) = d \leq \bar{d}$,

$$(9.15) \qquad\qquad B^{(k)} \;\subset\; \mathcal{N}_{\delta d}(A) \,.$$

Proof. We shall show the property for $k = 2$. This will show how the recursion works and the general result easily follows. We begin by noticing that if $x \in B(x_0, \delta)$, $y \in B(y_0, \delta)$, then $2x - y \in B(2x_0 - y_0, 3\delta)$. Thus

$$(9.16) \qquad (\mathcal{N}_{\delta d}(B))^{(1)} \subset \mathcal{N}_{3\delta d}((B)^{(1)}).$$

Applying First Reflection Lemma 9.9 with A' instead of A (where A' has been defined by a first application of the lemma, yielding a constant $\bar{d}$), A'' instead of A' and denoting by $\bar{d}^1$ the corresponding $\bar{d}$, we obtain that if $B \subset A''$ satisfies diam $B = d \le \bar{d}^1 \le \frac{\bar{d}}{4}$, then

$$(9.17) \qquad B^{(1)} \subset \mathcal{N}_{\delta d}(A').$$

For any $B \subset A''$, we set $B^1 = A' \cap \mathcal{N}_{\delta d}(B^{(1)})$. Relation (9.17) yields

$$B^{(1)} \subset \mathcal{N}_{\delta d}(A') \cap B^{(1)} \subset \mathcal{N}_{\delta d}(A' \cap \mathcal{N}_{\delta d}(B^{(1)})).$$

Thus

$$(9.18) \qquad B^{(1)} \subset \mathcal{N}_{\delta d}(B^1).$$

Applying this relation with B^1 instead of B:

$$(9.19) \qquad (B^1)^{(1)} \subset \mathcal{N}_{\delta d}((B^1)^1) \subset \mathcal{N}_{\delta d}(A).$$

We notice that by (9.1), diam$B^1 \le 4$diam$B \le \bar{d}$ (we take $\delta \le 1$). Thus by using, in this order, (9.18), (9.16) and (9.19) :

$$\begin{aligned} B^{(2)} &\subset (\mathcal{N}_{\delta d}(B^1))^{(1)} \\ &\subset \mathcal{N}_{3\delta d}((B^1)^{(1)}) \\ &\subset \mathcal{N}_{3\delta d}(\mathcal{N}_{\delta d}(A)) \\ &\subset \mathcal{N}_{4\delta d}(A). \end{aligned}$$

$\square$

2. Approximate tangent spaces.

A relevant consequence of Recursive Reflection Lemma 9.14 is the existence, at most points of A, of "approximate affine tangent spaces". For any real number α, we denote by $[\alpha]$ the last integer m such that $m \le \alpha$.

Lemma 9.20. *Let A be a regular α-set of $\mathbb{R}^N$. Then, for every $\varepsilon > 0$ there is a number $\overline{d} > 0$ and a subset A' of A with*

$$\mathcal{H}^\alpha(A \setminus A') < \varepsilon$$

such that for every $x \in A'$ and every positive number $r < \overline{d}$ one can find an affine subspace V of $\mathbb{R}^N$, with $x \in V$ and $\dim V = [\alpha]$ satisfying

(9.21) $A' \cap B_r(x) \subset \mathcal{N}_{\varepsilon r}(V \cap B_r(x)) \, .$

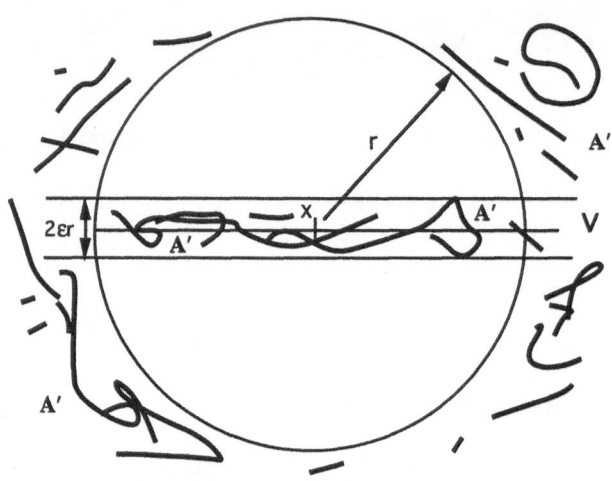

Figure 9.3-bis: Approximate tangent space (Lemma 9.20)

Proof. Preliminary explanation.
Before starting with the proof, let us give a sketchy argument to explain its main idea. Take for simplicity the case $\alpha = 1$ and assume that A is a subset of the plane. If at some point x of a regular subset A there is no tangent line, then we can find two other points of A, say y and z, arbitrarily close to x, such that x, y and z are not aligned at all. If we apply Recursive Reflection Lemma 9.14 to the three points, we can construct a net of points $y_1, y_2, \ldots, y_{n^k}$ of A which tends to "cover" the plane and remains close to A. Let us call εr the minimal distance between points in the net and take a ball with radius $\frac{\varepsilon r}{2}$ around each one of them. We know that the balls will contain in general a subset of A with length close to $2\varepsilon r$, because A is regular. Since we have a two-dimensional net, the ball $B(x, r)$ with radius r contains a number of

points of the net proportional to $\frac{r^2}{\varepsilon^2}$. Therefore the length of A inside $B(x,r)$ is proportional to $\frac{r^3}{\varepsilon}$. If ε is small enough, this contradicts the fact that the length of A inside $B(x)$ must be proportional to $2r$, the diameter of B when r is small. (See Figure 9.3.) Let us now develop the complete proof.

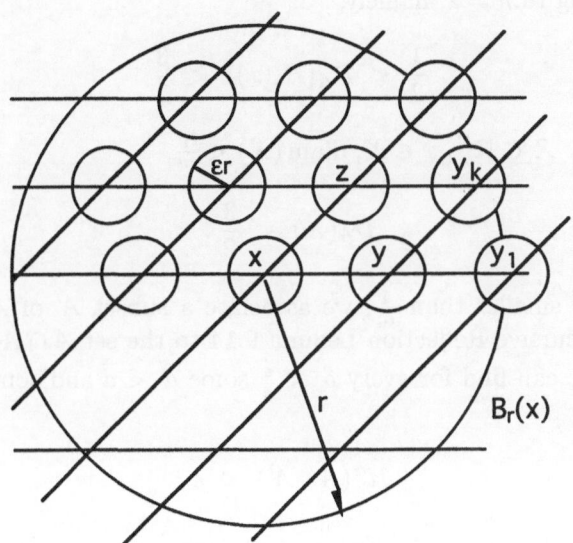

<u>Figure 9.3</u> : Net of points covering the plane and close to A.

Step 1.
If the lemma is not true, we can fix $\varepsilon > 0$ such that for every $\bar{d}$ and every A' satisfying $\mathcal{H}^\alpha(A \setminus A') < \varepsilon$, there is some $r < \bar{d}$ and some x in A' for which (9.21) does not hold whatever the choice of V is. Then, by using recursively the negation of (9.21), we can construct a finite sequence $x_0, x_1, \ldots, x_k$, with $k = [\alpha] + 1$, such that every x_i is in $A' \cap B_r(x)$ and (9.5) is satisfied for $\delta = \varepsilon r$, namely

$$\forall\, h = 1, 2, \ldots, k \ : \quad d(x_h, S_h) \geq \varepsilon r,$$

where S_h is the affine space spanned by $x_0, x_1, \ldots, x_{h-1}$. So, by First Geometric Lemma 9.4, for every choice of n, the reflected set $((A' \cap B_r(x))_{(n)})^{(k)}$ contains n^k points $y_1, y_2, \ldots, y_{n^k}$ at a distance larger or equal to εr from each other.

Step 2. Let us now specify the set A' and the number $\bar{d}'$ which will yield a contradiction. We fix an integer number m in such a way that

$$\mathcal{H}^\alpha(A \setminus A_{2,m}) < \frac{\varepsilon}{2} .$$

Recall that $A_{2,m}$ is defined as the set of the points x in $\mathbb{R}^N$ such that for every $r < \frac{1}{m}$ we have the particular instance of (8.26) and (8.27) corresponding to $n = 2$, namely,

$$(9.22) \qquad\qquad \frac{1}{2} < \mathcal{D}_A(B_r(x)) < \frac{3}{2}$$

and for every $X \subset \mathbb{R}^N$, $x \in X$, diam $(X) < \frac{1}{m}$,

$$(9.23) \qquad\qquad \mathcal{D}_A(X) < \frac{3}{2} .$$

With each $\bar{d}$ smaller than $\frac{1}{m}$, we associate a subset A' of $A \cap A_{2,m}$ by applying Recursive Reflection Lemma 9.14 to the set $A \cap A_{2,m}$ (instead of A). So we can find for every $\delta < \frac{\varepsilon}{4}$ some $\bar{d}' < \bar{d}$ and some subset A' such that

$$\mathcal{H}^\alpha(A \setminus A') < \varepsilon$$

and

$$(9.24) \qquad\qquad B^{(k)} \subset \mathcal{N}_{\delta d}(A_{2,m}) .$$

for every $B \subset A'$ with diam $(B) \le \bar{d}'$. A' and $\bar{d}$ being thus fixed, we choose x and r as specified in Step 1 of this proof. So the conclusions of Step 1 hold and in addition, by (9.2), for every choice of n we have

$$((A' \cap B_r(x))_{(n)})^{(k)} \subset (A' \cap B_r(x))^{(n+k)} .$$

Since also

$$\text{diam}\,(A' \cap B_r(x)) \le 2r$$

and $r \le \bar{d}$, we get from (9.24)

$$((A' \cap B_r(x))_{(n)})^{(k)} \subset \mathcal{N}_{\delta r}(A_{2,m}) .$$

Since $4\delta < \varepsilon$, by taking, for every $i = 1, 2, \ldots, n^k$, a point z_i in $A_{2,m}$ such that $|z_i - y_i| < \delta r$, we can find n^k points $(z_1, z_2, \ldots, z_{n^k})$ of $A_{2,m}$ which have a distance larger or equal to $\frac{\varepsilon}{2}r$ from each other and are contained in the δr-neighbourhood of $((A' \cap B_r(x))_{(n)})^{(k)}$. By using (9.3), we see that this set has a diameter less or equal to $2(3^k(2n+1)+\delta)r$ and contains x.

We can consider the n^k disjoint balls B_i centered at z_i and with radius $\frac{\varepsilon}{4}r$, thus contained in a $(\delta + \frac{\varepsilon}{4})r$-neighbourhood, B, of $((A' \cap B_r(x))_{(n)})^{(k)}$, with diameter less or equal to $2(3^k(2n+1) + \delta + \frac{\varepsilon}{4})r$. Since x and $z_1, z_2, \ldots z_{n^k}$ are in $A_{2,m}$, by (9.22) and (9.23) we have, for every $k, n, \bar{d}$ satisfying

$$(9.25) \qquad 2(3^k(2n+1) + \delta + \frac{\varepsilon}{4})\bar{d} < \frac{1}{m}$$

$$(9.26) \qquad \forall i : \quad \mathcal{D}_A(B_i) \geq \frac{1}{2}$$

$$(9.27) \qquad \mathcal{D}_A(B) \leq \frac{3}{2}.$$

From (9.26) we deduce

$$\mathcal{H}^\alpha(A \cap B) \geq \mathcal{H}^\alpha\left(A \cap \bigcup_{i=1}^{n^k} B_i\right) = \sum_{i=1}^{n^k} \mathcal{H}^\alpha(A \cap B_i)$$

$$\geq \sum_{i=1}^{n^k} c_\alpha \left(\frac{\varepsilon}{2}r\right)^\alpha \mathcal{D}_A(B_i) \geq \frac{1}{2} c_\alpha \left(\frac{\varepsilon}{2}r\right)^\alpha n^k .$$

On the other hand, (9.27) yields

$$\mathcal{H}^\alpha(A \cap B) = c_\alpha \mathcal{D}_A(B) \text{ diam} (B)^\alpha$$

$$\leq \frac{3}{2} 2^\alpha c_\alpha (3^k(2n+1) + \delta + \frac{\varepsilon}{4})^\alpha r^\alpha .$$

From the last two inequalities we finally deduce that

$$(\frac{\varepsilon}{2})^\alpha n^k \leq 3 \quad 2^\alpha (3^k(2n+1) + \delta + \frac{\varepsilon}{4})^\alpha,$$

provided $\bar{d}$ is chosen small enough to ensure that (9.25) is true. Since $\alpha < k$, we can find a convenient value of n such that the last inequality is false. This proves (9.21) by contradiction. $\quad\square$

We say that a set of sets $\mathcal{B}$ is a *Vitali almost covering* of a set A if there is a set $\tilde{A} \subset A$ such that $\mathcal{H}^\alpha(\tilde{A} \setminus A) = 0$ and $\mathcal{B}$ is a Vitali covering of $\tilde{A}$. Of course, this implies that $\mathcal{B}$ is an almost covering of A and Vitali Covering Lemma 7.14 can be applied to A and $\mathcal{B}$ as well as to $\tilde{A}$ and $\mathcal{B}$.

Corollary 9.28. *If A is a regular α-subset of $\mathbb{R}^N$, then for every positive number ε one can find a Vitali almost covering of A consisting of sets of the type $B_r(x) \cap \mathcal{N}_{\varepsilon r}(V)$, where x is a point of A and V is an affine subspace of $\mathbb{R}^N$, with $x \in V$ and $\dim V = [\alpha]$, such that $\mathcal{D}_A(B_r(x) \cap \mathcal{N}_{\varepsilon r}(V)) \geq 1 - \varepsilon$ and $\mathcal{D}_A(B_r(x) \setminus \mathcal{N}_{\varepsilon r}(V)) < \varepsilon$.*

Proof. Since A is regular, we can find by Lemma 8.28 for every $\varepsilon > 0$ a subset A_1 of A such that $\mathcal{H}^1(A \setminus A_1) < \varepsilon$ and for every $r < r_0$ small enough and every $x \in A_1$,

$$1 - \frac{\varepsilon}{2} < \mathcal{D}_A(B_r(x)) < 1 + \frac{\varepsilon}{2}.$$

Applying Lemma 9.20 to A_1 and $\frac{\varepsilon}{2}$ and thereafter Lemma 8.28, we find a subset A' of A_1 such that $\mathcal{H}^1(A \setminus A') < \varepsilon$ and for every $r < \min(r_0, \bar{d})$, affine subsets $V = V(x)$ such that

$$\mathcal{D}_{A'}(B_r(x)) = \mathcal{D}_{A'}(\mathcal{N}_{\varepsilon r}(V \cap B_r(x))).$$

In addition,

$$\mathcal{D}_{A'}(B_r(x)) \geq 1 - \frac{\varepsilon}{2}.$$

Thus $\mathcal{D}_A(B_r(x) \cap \mathcal{N}_{\varepsilon r}(V)) \geq 1 - \frac{\varepsilon}{2}$ and

$$\begin{aligned}
\mathcal{D}_A(B_r(x) \setminus \mathcal{N}_{\varepsilon r}(V)) &= \mathcal{D}_A(B_r(x)) - \mathcal{D}_A(B_r(x) \cap \mathcal{N}_{\varepsilon r}(V)) \\
&\leq 1 + \frac{\varepsilon}{2} - \mathcal{D}_{A'}(B_r(x) \cap \mathcal{N}_{\varepsilon r}(V)) \\
&\leq 1 + \frac{\varepsilon}{2} - (1 - \frac{\varepsilon}{2}) = \varepsilon.
\end{aligned}$$

Since we can apply again the preceding result to $A \setminus A'$ and $\frac{\varepsilon}{2}$, which still is regular, and so on, we get the thesis. $\square$

Lemma 9.29. *Let V be an affine subspace of $\mathbb{R}^N$ with $\dim V = [\alpha]$, B a ball of V with radius 1. Then if α is an integer, there exist for any $1 > \varepsilon > 0$ two real numbers $\varepsilon' > 0$ and $\lambda > 0$ such that for every x in $\overline{B}$ one can find an open covering $\mathcal{A}$ of $\mathcal{H}^\alpha(\mathcal{N}_{\varepsilon'}(V \cap (\overline{B} \setminus B')))$, where $B' = B(x, \varepsilon)$, satisfying*

$$(9.30) \qquad\qquad |\mathcal{A}|_\alpha \leq \lambda 2^\alpha c_\alpha.$$

If α is not an integer, the same result holds even if we let $B' = \emptyset$.

Proof. We have

$$\mathcal{H}^\alpha(V \cap (\overline{B} \setminus B(x, \varepsilon))) \leq c_\alpha(2^\alpha - \varepsilon^\alpha) ,$$

because $B \cap B'$ clearly contains a ball with diameter ε. So we can find an open covering $\mathcal{A}$ of $V \cap \overline{B} \setminus B'$ and a constant $\lambda < 1$ such that (9.30) holds.

Since $\overline{B} \setminus B(x, \varepsilon)$ is compact, $\mathcal{A}$ also covers a $2\varepsilon'$-neighbourhood of $V \cap (\overline{B} \setminus B')$ for some positive real number $\varepsilon'(x)$. Thus if $y \in B(x, \varepsilon')$, $\mathcal{A}$ also covers an ε'-neighbourhood of $V \cap (\overline{B} \setminus B(y, \varepsilon))$. Therefore, we may fix $\varepsilon'(y) = \varepsilon'(x)$ for $y \in B(x, \varepsilon')$. $\overline{B}$ being compact, we can cover it with finitely many such balls $B(x_i, \varepsilon'_i)$, with their associated λ_i. We finally fix $\varepsilon' = \min_i \varepsilon'_i$ and $\lambda = \max_i \lambda'_i$. $\quad\square$

The next result is a remarkable consequence of the existence almost everywhere of an "approximate tangent space", proved in Lemma 9.20. Let us write it in an informal way. From the existence, at almost every point x of a regular set A, of an approximate tangent space V, we shall conversely infer that there are points of A close to any point of V in the neighbourhood of x. In other terms, A is tangent to its tangent spaces. Of course, this is an important step towards the proof of the uniqueness of the tangent space at almost every point.

Lemma 9.31. *Let A be a regular α-set of $\mathbb{R}^N$. Then α must necessarily be an integer. In such a case, for every $\delta > 0$ one can find a number $\overline{d} > 0$ and a subset A' of A such that*

$$\mathcal{H}^\alpha(A \setminus A') < \delta$$

and that for every $x \in A'$ and every positive number $r < \overline{d}$ one can find an affine subspace V of $\mathbb{R}^N$, with $x \in V$ and $\dim V = \alpha$, such that

$$(9.21) \qquad A' \cap B_r(x) \subset \mathcal{N}_{\varepsilon r}(V \cap B_r(x))$$

and the converse relation

$$(9.32) \qquad V \cap B_r(x) \subset \mathcal{N}_{\varepsilon r}(A \cap B_r(x)) .$$

Proof. We begin by fixing $\varepsilon > 0$ and then ε' and $\lambda < 1$ such that the result of Lemma 9.29 holds. We then fix $A' \subset A$ such that $\mathcal{H}^\alpha(A \setminus A') < \delta$ and that

$$(9.33) \qquad A' \cap B_r(x) \subset \mathcal{N}_{\varepsilon' r}(V \cap B_r(x))$$

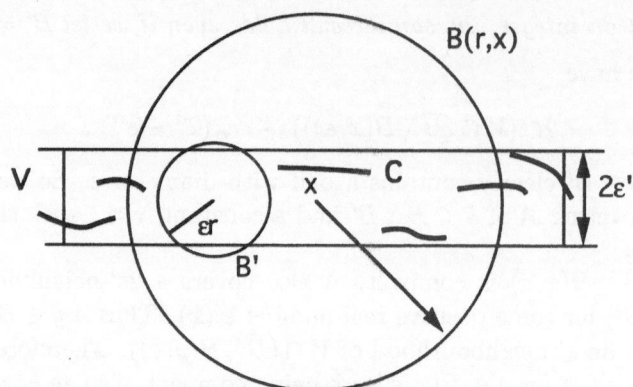

Figure 9.4 : Projection density (illustrates (9.36))

is satisfied (by applying Lemma 9.20) for any $r < \overline{d}$ and x in A'. Let us
denote for clarity by $V(x, r, \varepsilon')$ any of the affine spaces V such that (9.33)
is true. Consider the subset C of A' consisting of all points x such that
(9.32) is not satisfied for some of the affine spaces $V(x, r, \varepsilon')$, no matter
how small r is. In other terms

$$
\begin{aligned}
(9.34) \quad C = \{x, \forall r_0 < \overline{d},\ \exists r < r_0,\ \exists V(x, r, \varepsilon'), \\
V(x, r, \varepsilon') \cap B_r(x) \not\subset \mathcal{N}_{\varepsilon r}(A \cap B_r(x))\}.
\end{aligned}
$$

By Corollary 8.24, C is a regular α-set and we can in addition assume
(by taking a subset with slightly smaller measure) that x and r satisfy

$$
(9.35) \qquad \mathcal{D}_C^\alpha(B_r(x)) \geq 1 - \frac{1}{n} \, ,
$$

for some arbitrarily fixed value of n. By the definition (9.34) of C, we
can find a ball B' with radius εr, centered at a point of $V \cap B_r(x)$ such
that

$$
A \cap B' = \emptyset \, .
$$

By (9.33) this implies that

$$
(9.36) \qquad B_r(x) \cap A' \subset \mathcal{N}_{\varepsilon' r}(V \cap (B_r(x) \setminus B')) \, .
$$

What happens is now clear: The balls $B_r(x)$ considered in (9.34) form
a covering of C and we are allowed to "excise" from them a smaller B',
with a radius in a constant ratio ε to the radius of B, which does not
meet C. So, if the $B_r(x)$ are a covering of C which approximates the
Hausdorff measure of C, we shall get a contradiction if C does not have a
zero measure because the preceding situation implies (roughly speaking)
$\mathcal{H}^\alpha(C) \leq (1 - \varepsilon)\mathcal{H}^\alpha(C)$. Let us make this argument rigorous with the
help of Lemma 9.29.

The balls $B_r(x)$ considered in (9.34) are a Vitali covering of C. Consider a sequence of points $(x_k)_{k \in \mathbb{N}}$ and numbers $(r_k)_{k \in \mathbb{N}}$ in such a way that the balls $B_{r_k}(x_k)$ are mutually disjoint. By (9.35) we have

$$(9.37) \qquad 2^\alpha c_\alpha \sum_k r_k^\alpha \leq (1 - \frac{1}{n})^{-1} \mathcal{H}^\alpha(C),$$

so we have a bound on $|\cdot|_\alpha$ computed on any disjoint covering consisting of balls of the above type. Hence, we can apply Vitali Covering Lemma 7.14 and take the sequences x_k and r_k in such a way that the balls $B_{r_k}(x_k)$ form an almost covering of C.

We know that (9.36) is satisfied with $x = x_k$, $r = r_k$ and fix $B' = B'_k$ and $V = V_k$. Recall that by using Lemma 9.29 we have chosen ε' and $\lambda < 1$, only depending on ε, such that for every k we can find a covering $\mathcal{A}_k$ of $\mathcal{N}_{\varepsilon' r_k}(V_k \cap (B_{r_k}(x_k) \setminus B'_k))$ satisfying

$$(9.38) \qquad |\mathcal{A}_k|_\alpha \leq \lambda 2^\alpha c_\alpha r_k^\alpha.$$

Let $\mathcal{A} = \bigcup_k \mathcal{A}_k$. By (9.36) $\mathcal{A}$ turns out to be an almost covering of C. By (9.37) and (9.38) we have

$$c_\alpha |\mathcal{A}|_\alpha \leq c_\alpha \sum_k |\mathcal{A}_k|_\alpha \leq \lambda 2^\alpha c_\alpha \sum_k r_k^\alpha$$

$$\leq \lambda (1 - \frac{1}{n})^{-1} \mathcal{H}^\alpha(C).$$

Since $\rho(\mathcal{A}) \leq \sup_k r_k$, we can make $\rho(\mathcal{A})$ as small as we want, so from the above inequality we get at the limit (by Lemma 7.4)

$$(9.39) \qquad \mathcal{H}^\alpha(C) \leq \lambda (1 - \frac{1}{n})^{-1} \mathcal{H}^\alpha(C).$$

Therefore

$$(9.40) \qquad \left(1 - \lambda (1 - \frac{1}{n})^{-1}\right) \mathcal{H}^\alpha(C) \leq 0.$$

Since $\lambda < 1$ and the value of λ only depends on ε, while the integer n in (9.35) can be chosen to be arbitrarily large, we obtain from (9.40)

$$(9.41) \qquad \mathcal{H}^\alpha(C) = 0.$$

When α is not an integer, by the second assertion of Lemma 9.29, we can get (9.38) without the use of (9.36), or, in other words, with $B' = \emptyset$, and therefore without any restriction on the points of C. So we can take $C = A'$ from the beginning of the proof and we obtain $\mathcal{H}^\alpha(A') = 0$ instead of (9.41), so that finally $\mathcal{H}^\alpha(A) = 0$, in contradiction with the assumption that A is an α-set. $\square$

3. Existence of a tangent space.

Assume now that A is a regular α-set of $\mathbb{R}^N$ and let ε be a small enough positive number. By the results established in the preceding section, we know that we can find two $\mathcal{H}^\alpha$-measurable sets A' and A'' and a number $\bar{d} > 0$ in such a way that

$$(9.42) \qquad\qquad A' \subset A'' \subset A,$$

$$(9.43) \qquad\qquad \mathcal{H}^\alpha(A \setminus A') < \varepsilon.$$

The only condition on A'' is

$$(9.44) \qquad \forall\, x \in A'' , \, \forall\, r < \bar{d} \, : \quad \mathcal{D}_A^\alpha(B_r(x)) > \frac{1}{2}.$$

A' is a subset of A'', chosen by Lemma 9.31, satisfying the following relation.

$\forall\, x \in A', \forall\, r < \bar{d} \, , \, \exists\, V \subset \mathbb{R}^N$ (α-dimensional affine space),

$$(9.45) \qquad\qquad V \cap B_r(x) \subset \mathcal{N}_{\varepsilon r}(A'' \cap B_r(x))$$

$$(9.46) \qquad\qquad A' \cap B_r(x) \subset \mathcal{N}_{\varepsilon r}(V \cap B_r(x)) .$$

We fix $x \in A'$, $r < \frac{1}{2}\bar{d}$ and we choose V satisfying (9.45)-(9.46). We shall denote by P_V the orthogonal projection from $\mathbb{R}^N$ onto V.

Of course, the distinction between A, A', A'' is necessary, but it also is essentially contingent. The reader may well think "A", when he reads A, A' or A'' in the proofs below. This will not alter the main geometric meaning of the proofs. In other terms, since A, A', A'' are arbitrarily close to each other by (9.42)-(9.43), everything works as if A itself satisfied the three relations (9.44), (9.45) and (9.46).

The next lemma gives a key estimate in order to prove that at each point $x \in A$, the orthogonal projection $P_V(A)$ of A onto V roughly speaking covers V in a neighbourhood of x. This is the next step in our program, which is to prove that A looks like a smooth manifold at most of his points and therefore, locally, like an affine space. We shall give in the next lemma a precise account of what happens if the projection of A onto V has "holes" near x. We shall see that if such a "hole" $B_\rho(y)$ exists, then $(P_V)^{-1}(B_{2\rho}(y))$ must contain a big amount (in measure) of A. One can understand the subtle geometric argument by looking at Figure 9.5 below. We take a hole whose radius ρ is maximal and we can therefore choose some $z \in A''$ whose projection is close to the "hole" $B_\rho(y)$. Since A has a tangent space W at z, this tangent space must contain lines which are "almost orthogonal" to V. Thus many points $s_1, s_2, \ldots, s_n$ of A'' can be found in A'' to be close to W and their projections onto V nearly fall together. Using the lower density property of A (9.44), near each one of the points s_i, one will deduce that too many points of A lie in the backprojection of $A \cap P_{V^{-1}}(B_{2\rho} \setminus B_\rho)$ and get a contradiction. Let us now pass to the precise statement and proof.

Lemma 9.47. *Let y be a point of $V \cap B_{(1-2\sqrt{\varepsilon})r}(x)$ and ρ be a positive number such that*

$$(9.48) \qquad B_\rho(y) \cap P_V(A'' \cap B_r(x)) = \emptyset,$$

$$(9.49) \qquad B_{2\rho}(y) \cap P_V(A' \cap B_{2r}(x)) \neq \emptyset.$$

Then

$$(9.50) \qquad \mathcal{H}^\alpha(A \cap B_r(x) \cap P_V^{-1}(B_{4\rho}(y))) \geq \frac{2^\alpha}{8} c_\alpha \sqrt{\varepsilon^{-1}} \rho^\alpha.$$

Proof. From (9.45) and (9.48) we see that $\rho < \varepsilon r$. So $\varepsilon^{-1}\rho < \bar{d}$. Therefore if we choose z in $A' \cap B_{2r}(x) \cap P_V^{-1}(B_{2\rho}(y))$ (which is not empty by (9.49)), we have the analogies of (9.45)-(9.46) with x replaced by z and r by $\varepsilon^{-1}\rho$. So we can find an α-dimensional affine subspace of $\mathbb{R}^N$, W, such that

$$(9.51) \qquad W \cap B_{\varepsilon^{-1}\rho}(z) \subset \mathcal{N}_\rho(A'' \cap B_{\varepsilon^{-1}\rho}(z))$$

$$(9.52) \qquad A' \cap B_{\varepsilon^{-1}\rho}(z) \subset \mathcal{N}_\rho(W \cap B_{\varepsilon^{-1}\rho}(z)).$$

From (9.46) we see that $d(z, P_V(z)) < 2\varepsilon r$ (because $2r < \bar{d}$) and, subsequently, $d(z, y) \leq 2\varepsilon r + 2\rho \leq 4\varepsilon r$. Thus

$$B_{\sqrt{\varepsilon^{-1}}\rho}(z) \subset B_{r-2\rho}(x),$$

provided $\varepsilon < \frac{1}{36}$. Indeed, this implies $6\varepsilon r < \sqrt{\varepsilon} r$, and since $\rho < \varepsilon r$, we get $d(y, z) \leq 4\varepsilon r < \sqrt{\varepsilon} r - 2\rho$. Since y has by assumption a distance to the boundary of $B_r(x)$ larger than $2\sqrt{\varepsilon} r$, we deduce that the distance between z and the boundary of $B_r(x)$ is larger than $\sqrt{\varepsilon} r + 2\rho \geq \sqrt{\varepsilon^{-1}}\rho + 2\rho$. From (9.51) we deduce that

$$W \cap B_{\sqrt{\varepsilon^{-1}}\rho}(z) \subset (W \cap B_{\varepsilon^{-1}\rho}(z)) \cap B_{r-\rho}(x)$$
$$\subset \mathcal{N}_\rho(A'' \cap B_{\varepsilon^{-1}\rho}(z)) \cap B_{r-\rho}(x)$$
$$\subset \mathcal{N}_\rho(A'' \cap B_r(x)).$$

Therefore (9.48) implies

$$y \notin P_V(B_{\sqrt{\varepsilon^{-1}}\rho}(z) \cap W).$$

This relation means that the α-dimensional ball $B_{\sqrt{\varepsilon^{-1}}\rho}(z) \cap W$ contains a radial segment R which has a projection on V contained in the segment joining y with $P_V(z)$ and therefore contained in $B_{2\rho}(y)$.

R is a segment with length $\sqrt{\varepsilon^{-1}}\rho$, so it contains $\left[\frac{\sqrt{\varepsilon^{-1}}}{4}\right] + 1$ points t_i, $i = 1, 2, \ldots, \left[\frac{\sqrt{\varepsilon^{-1}}}{4}\right] + 1$, which have a distance larger or equal to 4ρ from each other. Since all of those points lie by construction in $B_{\varepsilon^{-1}\rho}(z) \cap W$, by (9.51), for every index i we can find a point s_i in A'' at a distance less or equal to ρ from t_i. The points s_i clearly belong to $P_V^{-1}(B_{3\rho}(y))$ and to $B_{r-\rho}(x)$. If $i \neq j$ then the distance between s_i and s_j is larger or equal to 2ρ, so the balls $B_\rho(s_i)$ are disjoint.

The balls $B_\rho(s_i)$ are contained in $B_r(x)$, and $P_V(B_\rho(s_i)) \subset B_{4\rho}(y)$. Since $\rho < \bar{d}$, by (9.44) we have

$$\mathcal{H}^\alpha(A \cap B_r(x) \cap P_V^{-1}(B_{4\rho}(y))) \geq \sum_i \mathcal{H}^\alpha(A \cap B_\rho(s_i))$$

$$\geq \frac{\sqrt{\varepsilon^{-1}}}{4} c_\alpha 2^\alpha \frac{1}{2} \rho^\alpha = \frac{2^\alpha}{8} c_\alpha \sqrt{\varepsilon^{-1}} \rho^\alpha. \quad \square$$

The estimate (9.50) of Lemma 9.47 suggests that when V is a tangent affine space to A at x, $P_V(A)$ cannot have inside $B_r(x)$ too many "holes"

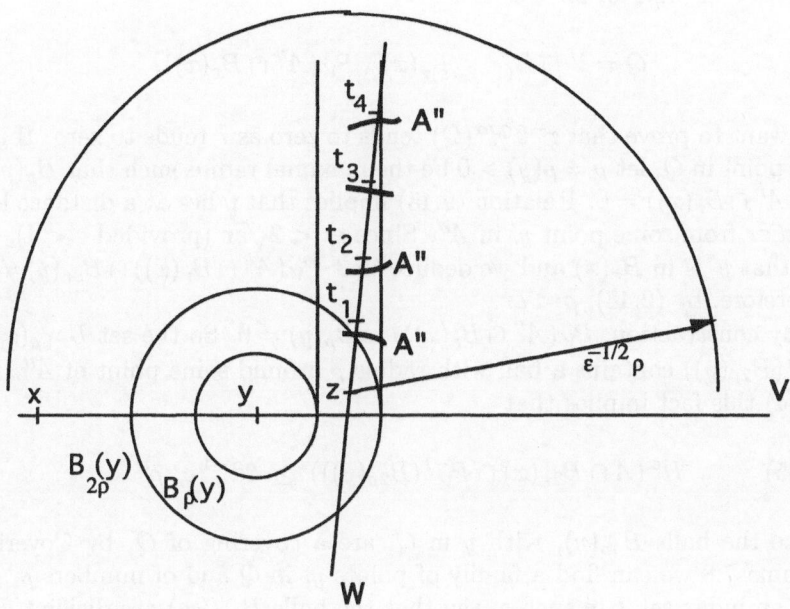

Figure 9.5 : Second projection density argument (Lemma 9.47)

$B_\rho(y)$. We shall now make a global use of this estimate by constructing a covering of all "holes" of $P_V(A) \cap B_r$ and thereafter proving that this set of holes has a density tending to zero when r tends to zero. If we make mentally $A = A' = A''$, the proof of the next lemma reduces to a few lines. Now, we have to work a bit more because Lemma 9.47 only states what happens to holes in the projection of A''. For every x in $\mathbb{R}^N$ we denote by $\mathcal{V}^\alpha(x)$ the set of all α-dimensional affine subspaces of $\mathbb{R}^N$ containing x.

Lemma 9.53 (Projection Density Lemma). *Let A be a regular α-set of $\mathbb{R}^N$. Then for almost every x in A,*

$$(9.54) \qquad \lim_{r \to 0} \left(\inf_{V \in \mathcal{V}^\alpha(x)} \frac{\mathcal{H}^\alpha(V \cap B_r(x) \setminus P_V(A \cap B_r(x)))}{r^\alpha} \right) = 0 .$$

Proof. Fix any $\varepsilon > 0$ and fix A', A'', $\bar{d}$ and V in such a way that the relations (9.42) to (9.46) hold. By using Proposition 6.26, we can choose for A'' a closed set.

Take $x \in A'$ and let $2r < \bar{d}$. Let

$$Q = V \cap B_{(1-2\sqrt{\varepsilon})r}(x) \setminus P_V(A'' \cap B_r(x)).$$

We want to prove that $r^{-\alpha} \mathcal{H}^\alpha(Q)$ tends to zero as r tends to zero. If y is any point in Q, let $\rho = \rho(y) > 0$ be the maximal radius such that $B_\rho(y) \cap P_V(A'' \cap B_r(x)) = \emptyset$. Relation (9.45) implies that y lies at a distance less than εr from some point y' in A''. Since $\varepsilon r < 2\sqrt{\varepsilon} r$ (provided $\varepsilon < 4$), we see that y' is in $B_r(x)$ and we deduce that $P_V(A'' \cap B_r(x)) \cap B_{\varepsilon r}(y) \neq \emptyset$. Therefore, by (9.48), $\rho < \varepsilon r$.

By construction, $P_V(A'' \cap B_r(x)) \cap \partial B_\rho(y) \neq \emptyset$. So the set $B_{r+\rho}(x) \cap P_V^{-1}(B_{2\rho}(y))$ contains a ball with radius ρ around some point of A''. By (9.44) this fact implies that

$$(9.55) \qquad \mathcal{H}^\alpha(A \cap B_{2r}(x) \cap P_V^{-1}(B_{2\rho}(y))) \geq 2^{\alpha-1} c_\alpha \rho^\alpha .$$

Since the balls $B_{4\rho}(y)$, with y in Q, are a covering of Q, by Covering Lemma 7.8 we can find a family of points y_i in Q and of numbers ρ_i for i in an index set I in such a way that the balls $B_{4\rho_i}(y_i)$ are disjoint and

$$(9.56) \qquad c_\alpha \sum_{i \in I} 8^\alpha \rho_i^\alpha \geq 5^{-\alpha} (\mathcal{H}^\alpha(Q) - \varepsilon) .$$

Let I_1 be the set of the indices i such that

$$B_{2\rho_i}(y_i) \cap P_V(A' \cap B_{2r}(x)) \neq \emptyset$$

and let $I_2 = I \setminus I_1$. We first show that the balls associated with I_2 do not weigh much. From (9.55),

$$\begin{aligned}
c_\alpha \sum_{i \in I_2} \rho_i^\alpha &\leq 2 \sum_{i \in I_2} \mathcal{H}^\alpha(A \cap B_{2r}(x) \cap P_V^{-1}(B_{2\rho_i}(y_i))) \\
&= 2 \sum_{i \in I_2} \mathcal{H}^\alpha((A \setminus A') \cap B_{2r}(x) \cap P_V^{-1}(B_{2\rho_i}(y_i))) \\
&\leq 2\mathcal{H}^\alpha((A \setminus A') \cap B_{2r}(x)) .
\end{aligned}$$

Since x is in A', this last term tends to zero faster than ρ^α, because Lemma 8.11 asserts that

$$\lim_{r \to 0} r^{-\alpha} \mathcal{H}^\alpha((A \setminus A') \cap B_{2r}(x)) = 0 .$$

If i is in I_1 we are in the situation of the previous lemma (9.47). So we have

$$c_\alpha \sum_{i \in I_1} \rho_i^\alpha \leq 8.2^{-\alpha} \sqrt{\varepsilon} \sum_{i \in I_1} \mathcal{H}^\alpha(A \cap B_r(x) \cap P_V^{-1}(B_{4\rho_i}(y_i)))$$

$$\leq 8.2^{-\alpha} \sqrt{\varepsilon} \, \mathcal{H}^\alpha(A \cap B_r(x)) \,.$$

By the assumption of regularity on A,

$$\lim_{r \to 0} r^{-\alpha} \mathcal{H}^\alpha(A \cap B_r(x)) = 2^\alpha c_\alpha.$$

Relation (9.56) and the arbitrariness of ε show that the above relations allow us to make $r^{-\alpha} \mathcal{H}^\alpha(Q)$ as small as we want by taking r small enough. This proves (9.54). □

Let us summarize our progress towards the existence of a tangent plane at almost all points of a regular set. From Lemmas 9.31 and 9.53 we immediately deduce the next proposition.

Proposition 9.57. *Let α be a positive real number and A be an α-set of $\mathbb{R}^N$. Then A is $\mathcal{H}^\alpha$-regular if and only if α is an integer and at almost every point x of A,*

$$(9.58) \qquad \lim_{r \to 0} \left(\sup_{V \in \mathcal{V}^\alpha(x)} \frac{\mathcal{H}^\alpha(P_V(A \cap B_r(x)))}{c_\alpha(2r)^\alpha} \right) = 1 \,.$$

Note that, in the last two statements, the estimates (9.54) and (9.58), involving extremal values of functions of V in $\mathcal{V}^\alpha(x)$, actually hold as simple limits, without the "sup" or the "inf", if we replace in them "V" by "$V(x,r)$", where $V(x,r)$ satisfies (9.45) and (9.46).

A natural question at this point is to ask whether we can find a limit space of $V(x,r)$ when r tends to zero. This will be based on a rather intuitive result, which is illustrated in Figure 9.6 below: Since the projection of an affine space V onto another affine space $\overline{V}$, with an angle φ to the first, reduces the Hausdorff measure by a factor $(\cos\varphi)^\alpha$, it is clear that A cannot be close to both spaces and have a projection with local density 1. In order to make the intuition correct, we fix $\overline{\varepsilon} > 0$ and $A' \subset A$ such that $\mathcal{H}^\alpha(A \setminus A') < \overline{\varepsilon}$ and (9.21), namely

$$A' \cap B_r(x) \subset \mathcal{N}_{\varepsilon r}(V(x,r) \cap B_r(x)),$$

holds for every $r < \bar{d}$ in correspondence of every $\varepsilon < \bar{\varepsilon}$, under a suitable choice of $\bar{d}$ depending on the value of ε. (We can easily determine such a set A' by taking the intersection of the sets A'_n obtained in Lemma 9.20 for $\varepsilon = 2^{-n}\bar{\varepsilon}$.) We can also assume that A'_n, and therefore A', satisfies the relations (9.42)-(9.46).

We fix for every x in A' two infinitesimal sequences $(\varepsilon_n)_{n \in \mathbb{N}}$ and $(\rho_n)_{n \in \mathbb{N}}$ such that for every n there is a space $V = V(x, \rho_n)$ satisfying (9.21) for $r = \rho_n$ and $\varepsilon = \varepsilon_n$.

Lemma 9.59. *For almost every x in A' the sequence of affine spaces $(V(x, \rho_n))_{n \in \mathbb{N}}$ is a Cauchy sequence.*

Proof. If the statement is not true, we can find a subset X of A' with positive measure and an angle $0 < \varphi < \frac{\pi}{2}$ such that if x is in X there are two numbers r_1 and r_2 (elements of the sequence $(r_n)_{n \in \mathbb{N}}$), arbitrarily small, such that $V(x, r_1)$ makes an angle larger than 2φ with $V(x, r_2)$. Let $\bar{x}$ be a point of X, let $\bar{r}$ small enough be fixed and let $\overline{V} \in \mathcal{V}^\alpha(\bar{x})$ satisfy (9.45) and (9.46). We may assume that

$$(9.60) \qquad\qquad |\mathcal{D}_X^\alpha(B_r(\bar{x})) - 1| < \varepsilon,$$

$$(9.61) \qquad\qquad |\mathcal{D}_{A \backslash X}^\alpha(B_r(\bar{x}))| < \varepsilon,$$

$$(9.62) \qquad\qquad \left| \frac{\mathcal{H}^\alpha(P_V(A \cap B_r(\bar{x})))}{c_\alpha (2\bar{r})^\alpha} - 1 \right| < \varepsilon,$$

in correspondence with a suitably small value of ε, as we can do by using the regularity assumption on A, the density relation (8.19) and (9.58). Then, by definition of X, we can find a Vitali covering of $X \cap B_{\bar{r}}(\bar{x})$ consisting of balls $B_r(x)$, with x in X, such that the space $V(x, r)$ makes an angle larger or equal than φ with $\overline{V}$ and $\varepsilon(r) \leq \varepsilon$. By the regularity of X we can also assume that for almost every x and r as above we have

$$(9.63) \qquad\qquad |\mathcal{D}_X^\alpha(B_r(x)) - 1| < \varepsilon .$$

Since $X \subset A'$ for every x and r we have from (9.21)

$$X \cap B_r(x) \subset \mathcal{N}_{\varepsilon r}(V(x,r) \cap B_r(x))$$

and therefore

$$P_{\overline{V}}(X \cap B_r(x)) \subset \mathcal{N}_{\varepsilon r}(B_{r \cos \varphi}(P_{\overline{V}}(x))).$$

Since the angle between $V(x,r)$ and $\overline{V}$ is larger or equal to φ, the Lipschitz constant of the projection of V onto $\overline{V}$ is less than $\cos \varphi$. So we deduce from the last inclusion that

$$(9.64) \qquad \mathcal{H}^{\alpha}(P_{\overline{V}}(X \cap B_r(x)) \le c_{\alpha}\,(\varepsilon + \cos\,\varphi)^{\alpha}\,(2r)^{\alpha}\,.$$

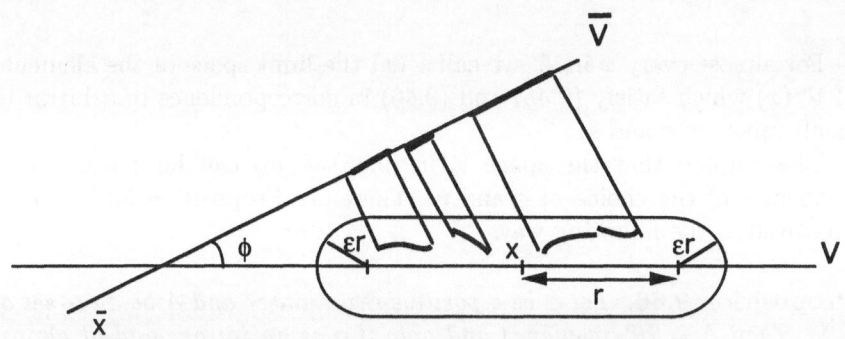

Figure 9.6 : continuity of tangent spaces (Lemma 9.59)

By Vitali Covering Lemma 7.14 we cover almost all of the set $X \cap B_r(\overline{x})$ with a family of disjoint balls $B_{r_i}(x_i)$ of the above type, contained in $B_r(\overline{x})$.

So from (9.63) and (9.60) we have

$$
\begin{aligned}
(9.65) \qquad c_{\alpha} \sum_i (2r_i)^{\alpha} &\le (1-\varepsilon)^{-1} \sum_i \mathcal{H}^{\alpha}(X \cap B_{r_i}(x_i)) \\
&\le (1-\varepsilon)^{-1}\,\mathcal{H}^{\alpha}(X \cap B_{\overline{r}}(\overline{x})) \\
&\le \frac{1+\varepsilon}{1-\varepsilon}\,c_{\alpha}(2\overline{r})^{\alpha}\,.
\end{aligned}
$$

Collecting the relations (9.61), (9.64) and (9.65) we get

$$\frac{\mathcal{H}^\alpha(P_{\overline{V}}(A \cap B_{\overline{r}}(\overline{x})))}{c_\alpha(2\overline{r})^\alpha} \leq$$

$$\leq \frac{\mathcal{H}^\alpha(P_{\overline{V}}((A \setminus X) \cap B_{\overline{r}}(\overline{x})))}{c_\alpha(2\overline{r})^\alpha} + \frac{\mathcal{H}^\alpha(P_{\overline{V}}(X \cap B_{\overline{r}}(\overline{x})))}{c_\alpha(2\overline{r})^\alpha}$$

$$\leq \varepsilon + c_\alpha^{-1}(2\overline{r})^{-\alpha} \sum_i \mathcal{H}^\alpha(P_{\overline{V}}(X \cap B_{r_i}(x_i)))$$

$$\leq \varepsilon + (2\overline{r})^{-\alpha} \sum_i (\varepsilon + \cos\varphi)^\alpha (2r_i)^\alpha$$

$$\leq \varepsilon + \frac{1+\varepsilon}{1-\varepsilon} (\varepsilon + \cos\varphi)^\alpha < 1 - \varepsilon \,,$$

provided ε has been chosen small enough. So we get a contradiction with Relation (9.62). $\square$

For almost every x in A' we call $V(x)$ the limit space of the elements of $\mathcal{V}^\alpha(x)$ which satisfy (9.45) and (9.46) in correspondence of arbitrarily small values of r and ε.

This implies that the space V in (9.45)-(9.46) can be chosen independently of the choice of ε and r. Therefore Proposition 9.57 can be improved in the following way.

Proposition 9.66. *Let α be a positive real number and A be an α-set of $\mathbb{R}^N$. Then A is $\mathcal{H}^\alpha$-regular if and only if α is an integer and at almost every $x \in A$,*

$$(9.67) \qquad \sup_{V \in \mathcal{V}^\alpha(x)} \left(\lim_{r \to 0} \frac{\mathcal{H}^\alpha(P_V(A \cap B_r(x)))}{c_\alpha(2r)^\alpha} \right) = 1 \,.$$

In addition, the "sup" is attained at a single space $V(x)$ for almost every $x \in A$.

A further property of the space $V(x)$ is that, near x, A is entirely contained in an "angular neighbourhood" of V. We say that $C = C(x, V, \varphi)$ is an angular neighbourhood of V if it contains all the points y such that $y - x$ makes an angle with V smaller than a certain value φ.

Proposition 9.68. *Let x be a point of A' such that the limit space $V(x)$ exists. Then if $C(x, V, \varphi)$ is any angular neighbourhood of V and if r is suitably small in dependence of φ,*

(9.69) $A' \cap B_r(x) \subset C$.

Consequently, we also have for almost every x in A'

(9.70) $d^\alpha_{A \setminus C}(x) = 0$.

Proof. If for r arbitrarily small we can find a point $y_r \in (A \setminus C) \cap \partial B_r(x)$ we see that a space V which satisfies (9.21) for a value of ε suitably small must make an angle larger or equal to $\frac{\varphi}{2}$ with $V(x)$. This contradicts the definition of $V(x)$. (See Figure 9.7.) $\square$

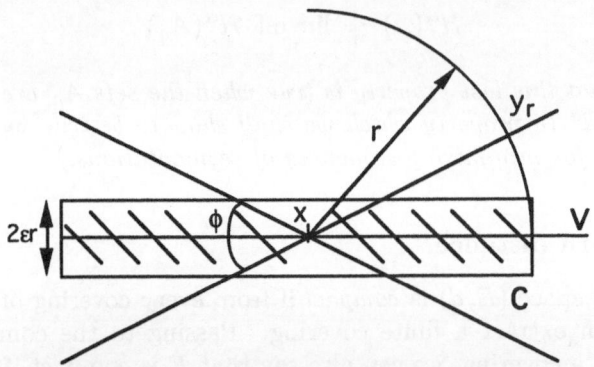

<u>Figure 9.7</u> : conic density (Proposition 9.68)

It is clear that we can define $V(x)$ for almost every x in A, modulo an arbitrary choice of A'. Then the formulas (9.67) and (9.70), which do not involve A', must hold. Moreover, both formulas imply the uniqueness of $V(x)$ for almost every x. Indeed, if $V \neq V(x)$, we can take C in such a way that

$$\forall r > 0 : \quad \frac{\mathcal{H}^\alpha(P_V(C \cap B_r(x)))}{c_\alpha(2r)^\alpha} < 1 .$$

We shall call the space $V(x)$ which has been uniquely characterized by Propositions 9.66 or 9.68 for almost every x in A, the *tangent space* to A at point x.

Chapter 10

SEMICONTINUITY PROPERTIES
OF THE HAUSDORFF MEASURE

In this chapter, we first define a natural metric on the set of the closed sets of a metric space E: the Hausdorff distance. For this distance, we prove that the set of closed subsets of a compact set is compact. It would be extremely convenient to have the following continuity property: If a sequence of sets A_n converges for the Hausdorff distance towards a set A, then the Hausdorff measures of the A_n also converge to the Hausdorff measure of A. We give examples which show that this is in general not true. However, simple and useful conditions can be given on the sequence A_n in order that the Hausdorff measure is lower semicontinuous, that is,

$$\mathcal{H}^\alpha(A) \leq \liminf_n \mathcal{H}^\alpha(A_n).$$

We show that this last property is true when the sets A_n are "uniformly concentrated" (a property which we shall show to be true in Chapter 15 of this book for minimizing sequences of segmentations.)

1. Hausdorff distance.

A metric space (E, d) is *compact* if from every covering of E by open sets one can extract a finite covering. Passing to the complementary sets of such a covering, we can also say that E is compact if and only if from every family of compact sets whose intersection is empty, a finite subfamily can be extracted for which the same property holds. One checks easily that a metric space is compact if and only if every sequence in E has a converging subsequence. We say that a metric space E is *precompact* if it has for every $\varepsilon > 0$ an ε-*net*, that is, a finite set N such that every point in E has a distance less than ε to some point of N. In other terms, the balls $B(x, \varepsilon), x \in N$ are a finite covering of E. If E is precompact, by a diagonal selection argument one can easily show that every sequence in E has a Cauchy subsequence. If every Cauchy sequence of a metric space E has a limit in E, then we say that E is *complete*. Therefore, if E is precompact and complete, it is compact: every sequence has a converging subsequence. The converse implication is also true.

We shall use the preceding framework to discuss the metric properties of the set $\mathcal{P}(E)$ of all subsets of E. We consider for any pair A and B of subsets of a given metric space E the Hausdorff semidistance

$$\mathbf{d}(A, B) = \max \left(\sup_{x \in A} d(x, B), \sup_{x \in B} d(x, A) \right)$$

where $d(x, A)$ denotes, as usually, the distance of the point x to the set A,

$$d(x, A) = \inf_{y \in A} d(x, y).$$

We say that $A \in \mathcal{P}(E)$ is the Hausdorff limit of a sequence of sets $(A_n)_{n \in \mathbb{N}}$ if A is closed and $\mathbf{d}(A, A_n) \to 0$. The Hausdorff limit is uniquely determined because $\mathbf{d}$ is a distance between closed sets. We point out that if E is compact every sequence of sets of E has a converging subsequence, as follows from the next proposition.

Proposition 10.1. *If E is compact, then $(\mathcal{P}(E), \mathbf{d})$ is compact.*

Proof. We begin by noticing that since E is precompact, $(\mathcal{P}(E), \mathbf{d})$ is precompact. Indeed, fix $\varepsilon > 0$ and let N be an ε-net of E. Then $\mathcal{P}(N)$ is clearly an ε-net for $(\mathcal{P}(E), \mathbf{d})$.

Therefore, every sequence $(A_n)_{n \in \mathbb{N}}$ has a Cauchy subsequence for the Hausdorff metric $\mathbf{d}$. Let us show that this sequence converges or, in other terms, that $\mathcal{P}(E)$ is complete. Set for every n

$$B_n = \overline{\bigcup_{k \geq n} A_k}$$

and

$$(10.2) \qquad A = \bigcap_n B_n .$$

Since $(A_n)_{n \in \mathbb{N}}$ is a Cauchy sequence, for every $\varepsilon > 0$ we can find n such that A_k is contained in the ε-neighbourhood of A_n for every $k \geq n$. So B_n is contained in the closed ε-neighbourhood of A_n and, since on the other hand $A_n \subset B_n$, we have that

$$\mathbf{d}(A_n, B_n) \leq \varepsilon .$$

By the arbitrariness of ε this means that

$$(10.3) \qquad \mathbf{d}(A_n, B_n) \to 0 .$$

The sequence $(B_n)_{n \in \mathbb{N}}$ being decreasing by inclusion, $A \subset B_n$ for every value of n. E being a compact space, if we take any open neighbourhood N of A, the sets B_n also are compact and $(B_n \setminus N)_{n \in \mathbb{N}}$ is a sequence of compact sets with an empty intersection. So N has to contain B_n for large n. This proves that

$$\mathbf{d}(B_n, A) \rightarrow 0$$

and finally by (10.3)

$$\mathbf{d}(A_n, A) \rightarrow 0 . \qquad \square$$

2. Counterexamples to the semicontinuity of the Hausdorff measure for the Hausdorff metric.

We have seen in this way that if E is a compact space then every sequence of subsets of E has a Cauchy subsequence for $\mathbf{d}$, and this subsequence has a Hausdorff limit. We are interested in finding conditions on a sequence of sets, which make the Hausdorff measure $\mathcal{H}^\alpha$ upper or lower semicontinuous. Note that such a property is clearly false in general. Let us take $E = [0, 1]$ and construct A_n by dividing $[0, 1]$ into n^2 equal subintervals and by taking the union of the intervals which correspond to the integers multiple of n. We see that

$$(10.4) \qquad \mathbf{d}(A_n, [0, 1]) = \frac{n-1}{n^2} \rightarrow 0$$

and

$$\mathcal{H}^1(A_n) = \frac{1}{n} \rightarrow 0 .$$

So we can find a sequence of sets which tends for $\mathbf{d}$ to the whole interval and whose measures tend to zero. Since $\mathcal{H}^1([0, 1]) = \mathcal{L}_1([0, 1]) = 1$, we see that $\mathcal{H}^1$ is not lower semicontinuous with respect to $\mathbf{d}$ along the sequence $(A_n)_{n \in \mathbb{N}}$.

Let us consider another example. We take the sets B_n obtained by fixing an integer k which does not depend on n, by dividing $[0, 1]$ into n subintervals, and by making the union of the ones corresponding to a number which is not a multiple of k. Again we have

$$\mathbf{d}(B_n, [0, 1]) \rightarrow 0$$

and on the other hand

(10.5) $$\mathcal{H}^1(B_n) \;\to\; \frac{k-1}{k} < 1 \,.$$

We have considered this second example in order to suggest a connection between the semicontinuity of $\mathcal{H}^\alpha$ and the density properties of the B_n. In the above example, every given measurable set A satisfies

(10.6) $$\lim_n \mathcal{D}_{B_n}(A) \;=\; \frac{k-1}{k}.$$

Therefore (10.6) is not enough to have the lower semicontinuity of $\mathcal{H}^1$ although the lack of semicontinuity is substantially reduced compared with the previous example. Actually one finds the same constant $\frac{k-1}{k}$ in (10.5) and (10.6) and the density property (10.6) produces the difference between (10.4) and (10.5). We shall see in the following that density assumptions related to (10.6) can guarantee a lower semicontinuity result. Simple counterexamples to the upper semicontinuity of Hausdorff measure can be seen by taking any $\alpha < 1$ and any nonempty closed subset A of $[0,1]$ with $\mathcal{H}^\alpha(A) < +\infty$. Then

$$\mathcal{N}_{\frac{1}{n}}(A) \;\to\; A \quad \text{by} \quad \mathbf{d}$$

and since, for every integer n, $\mathcal{N}_{\frac{1}{n}}(A)$ is a nonempty open set, we have $\mathcal{H}^1(\mathcal{N}_{\frac{1}{n}}(A)) > 0$ and so $\mathcal{H}^\alpha(\mathcal{N}_{\frac{1}{n}}(A)) = +\infty$. (It is easy to check directly from the definition of the Hausdorff measure (6.1) that if $\mathcal{H}^\alpha(A) > 0$, then $\mathcal{H}^\beta(A) = +\infty$ for any $\beta < \alpha$.)

We can also construct a counterexample for the measure $\mathcal{H}^1$, but we have to consider subsets of $\mathbb{R}$ which are not contained in $[0,1]$. If we take in the above example $A = \mathbb{N}$, we see that $\mathcal{H}^1(A) = 0$ and, for every n, $\mathcal{H}^1(\mathcal{N}_{\frac{1}{n}}(A)) = +\infty$.

3. Upper semicontinuity in $[0,1]$.

However, if we are concerned with subsets of $[0,1]$, the upper semicontinuity property of $\mathcal{H}^1$ is ensured. Note that only the finiteness of the measure of $[0,1]$ is going to play a role in the following argument.

Proposition 10.7. *Let* $X = [0,1]$. *If* $(A_n)_{n \in \mathbb{N}}$ *is a sequence of measurable subsets of* X *and* A *is the Hausdorff limit of* $(A_n)_{n \in \mathbb{N}}$ *, then*

(10.8) $$\limsup_n \mathcal{H}^1(A_n) \;\leq\; \mathcal{H}^1(A) \,.$$

Proof. If we denote by f_n the positive part of the difference between the characteristic function of A_n and that of A,

$$f_n = (1_{A_n} - 1_A)^+ ,$$

by the definition of the Hausdorff limit, f_n converges to zero pointwise in X. Since the functions f_n lay below the constant function 1 in X, by the Lebesgue Theorem we see that

$$\int f_n \, d\mathcal{H}^1 \to 0 ,$$

namely

$$\mathcal{H}^1(A_n \setminus A) \to 0 .$$

Using

$$\mathcal{H}^1(A_n) \leq \mathcal{H}^1(A_n \setminus A) + \mathcal{H}^1(A_n \cap A)$$
$$\leq \mathcal{H}^1(A_n \setminus A) + \mathcal{H}^1(A),$$

(10.8) follows.   □

4. Lower semicontinuity properties of the Hausdorff measure on uniformly concentrated sets.

Coming back to the lower semicontinuity case, we consider the following property. If $(A_n)_{n \in \mathbb{N}}$ is a sequence of subsets of E which converges for **d** to a subset A, we say that $(A_n)_{n \in \mathbb{N}}$ is *uniformly concentrated* (or has the *Uniform Concentration Property*) if for every $\varepsilon > 0$ we can find a Vitali approximate covering $\mathcal{B}$ of A such that

(10.9) $\forall\, B \in \mathcal{B} : \quad \limsup\limits_{n} \mathcal{D}^{\alpha}_{A_n}(B) > 1 - \varepsilon .$

(Compare with the relation (10.6) discussed above!)

Thus, in order to show that $(A_n)_{n \in \mathbb{N}}$ is uniformly concentrated, we have to find some ε' and a Vitali covering of A, $\hat{\mathcal{B}}$, and then a set of subsets of E, $\mathcal{B}$ such that (10.9) holds and also (7.13), namely

$$\forall\, \hat{B} \subset \hat{\mathcal{B}} \; \exists\, B \in \mathcal{B} \quad \text{such that} \quad B \subset \hat{B} \quad \text{and}$$
$$\operatorname{diam}(B) \geq \varepsilon' \operatorname{diam}(\hat{B}) .$$

When $\hat{\mathcal{B}}$ only contains compact sets, then we can obtain the Uniform Concentration Property provided that, for every $\hat{B}$ in $\hat{\mathcal{B}}$, we can find a subset B_n, for n large, in such a way that (7.13) holds for $B = B_n$ and $\mathcal{D}_{A_n}(B_n) > 1 - \varepsilon$. The fact that B_n changes with n is not a real difficulty, under the substitution of $(A_n)_{n \in \mathbb{N}}$ by a subsequence:

Proposition 10.10. *Let $(A_n)_{n\in\mathbb{N}}$ be a sequence of subsets of E converging to A for d and assume that, for every $\varepsilon > 0$, we can find a countable Vitali covering $\hat{\mathcal{B}}$ of A and a constant $c > 0$ such that*

(10.11)
$$\forall\,\hat{B} \in \hat{\mathcal{B}}\,,\; \exists\,\overline{n} \in \mathbb{N}\quad \forall\, n \geq \overline{n}\,,\; \exists\, B_n \in \hat{\mathcal{B}},$$
$$\operatorname{diam} B_n \geq c\operatorname{diam}\hat{B} \text{ and } \mathcal{D}^\alpha_{A_n}(B_n) > 1 - \varepsilon\,.$$

Assume also that every set in $\hat{\mathcal{B}}$ is compact. Then $(A_n)_{n\in\mathbb{N}}$ has a uniformly concentrated subsequence.

Proof. For every $\hat{B}$ in $\hat{\mathcal{B}}$ the sequence of sets $(B_n)_{n\in\mathbb{N}}$ (defined for n large) has a converging subsequence for d, since $\hat{B}$ is compact. So, if we replace $(A_n)_{n\in\mathbb{N}}$ by a suitable subsequence, we can assume that $(B_n)_{n\in\mathbb{N}}$ converges. Since we have taken $\hat{\mathcal{B}}$ countable, we can assume that such a property holds simultaneously for every $\hat{B}$ in $\hat{\mathcal{B}}$ by a diagonal selection procedure. Let B be a neighbourhood of the Hausdorff limit of $(B_n)_{n\in\mathbb{N}}$. We can choose B in such a way that

(10.12)
$$(1 - \varepsilon)\, \operatorname{diam} B \leq \lim_n \operatorname{diam} B_n\,.$$

By (10.12) and (10.11) we get, for n large enough,

(10.13)
$$\mathcal{D}^\alpha_{A_n}(B) = \frac{\mathcal{H}^\alpha(A_n \cap B)}{c_\alpha(\operatorname{diam} B)^\alpha} \geq (1-\varepsilon)^\alpha \frac{\mathcal{H}^\alpha(A_n \cap B_n)}{c_\alpha(\operatorname{diam} B_n)^\alpha}$$
$$= (1-\varepsilon)^\alpha\, \mathcal{D}^\alpha_{A_n}(B_n) \geq (1-\varepsilon)^{\alpha+1}\,.$$

The set of the sets B, obtained in this way, is clearly a Vitali almost covering of A. Indeed we have, for instance, $B \subset \hat{B}^*$ (where we denote, as in Chapter 7, by $\hat{B}^*$ the 2diam(B)-neighbourhood of B). Thus

$$\operatorname{diam} B \geq \lim_n \operatorname{diam} B_n \geq c\,\operatorname{diam}\hat{B} \geq c\,5^{-1}\operatorname{diam}\hat{B}^*\,.$$

Moreover, $\hat{\mathcal{B}}^*$ still is a Vitali covering of A. Therefore, by the arbitrariness of ε, we get the thesis. $\quad\square$

Remark.
(i) The assumptions in the above proposition are general enough for the applications in the other chapters. However, they can easily be weakened. We first note that the assumption that every $\hat{B}$ in $\hat{\mathcal{B}}$ is compact can be

clearly substituted by the condition that all the sets B_n are contained in a same relatively compact subset of $\hat{B}$. Then the assumption that $\hat{B}$ is countable (which has been made to the aim of replacing $(A_n)_{n\in\mathbb{N}}$ with a subsequence rather than with a more general subnet) is not a real restriction since it can be forced, provided

$$\liminf_n \mathcal{H}^\alpha(A_n) < +\infty ,$$

which is the only case in which we shall be interested in using the Uniform Concentration Property.

(ii) For every integer k, by Lemma 7.1, we can find a subset $\mathcal{B}_k$ of $\hat{B}$ consisting of mutually disjoint sets such that $\rho(\mathcal{B}_k) < \frac{1}{k}$ and $\mathcal{B}_k^*$ is a covering of A. Then $(\mathcal{B}_k^*)_{k\in\mathbb{N}}$ is another Vitali covering of A which can replace $\hat{B}$ in the above proposition. Moreover, every $\mathcal{B}_k$ is at most countable because it contains disjoint sets and each one of them has a positive density with respect to some set A_n of finite measure.

As stated in the next (main) result, the Uniform Concentration Property is enough to ensure the lower semicontinuity of the Hausdorff measure with respect to the Hausdorff metric of sets.

Theorem 10.14. *Assume that a sequence of subsets A_n tends to a set A for $\mathbf{d}$ and that $(A_n)_{n\in\mathbb{N}}$ is uniformly concentrated . Then*

$$(10.15) \qquad \mathcal{H}^\alpha(A) \leq \liminf_n \mathcal{H}^\alpha(A_n) .$$

Proof. Fix $\varepsilon > 0$ and let $\mathcal{B}$ be such that (10.9) holds. Fix any real number $b < \mathcal{H}^\alpha(A)$. By Vitali Covering Lemma 7.14 we can find a finite set $\mathcal{A} \subset \mathcal{B}$ such that $\mathcal{A}$ contains mutually disjoint sets and

$$(10.16) \qquad c_\alpha |\mathcal{A}|_\alpha > b .$$

By (10.9) we see that we can find a number $\bar{n}$ such that

$$(10.17) \qquad \forall\, B \in \mathcal{A} ,\, \forall\, n \geq \bar{n} : \quad \mathcal{D}^\alpha_{A_n}(B) > 1 - \varepsilon .$$

By (10.16), (10.17) we have for $n \geq \bar{n}$

$$(10.18) \qquad \begin{aligned} \mathcal{H}^\alpha(A_n) &\geq \mathcal{H}^\alpha\!\left(A_n \cap \bigcup \mathcal{A}\right) = \sum_{B\in\mathcal{A}} \mathcal{H}^\alpha(A_n \cap B) \\ &\geq (1-\varepsilon)\, c_\alpha |\mathcal{A}|_\alpha > (1-\varepsilon) b . \end{aligned}$$

By the arbitrariness of b and ε we get (10.15). $\qquad\square$

5. Sequences of compact connected 1-sets.

Theorem 10.19 (Golab's Theorem). *Let* $(A_n)_{n\in\mathbb{N}}$, *be a sequence of connected subsets of* $\mathbb{R}^N$ *which converges to a set* A *for* **d**. *Then*

$$(10.20) \qquad \mathcal{H}^1(A) \leq \liminf_n \mathcal{H}^1(A_n) \,.$$

Proof. Relation (10.20) will follow from the semicontinuity theorem (10.14) if we show that $(A_n)_{n\in\mathbb{N}}$ satisfies the assumptions of Proposition 10.10. To this aim, let $\hat{B}$ consist of the balls centered at a point of A and which do not contain A_n for n large. Fix $\hat{B} \in \hat{B}$ and $1 > \varepsilon > 0$. For n large enough, if x is the center of $\hat{B}$ and r its radius we can find x_1 in A_n with

$$d(x, x_1) \leq \frac{\varepsilon}{4} r$$

and, since A_n is connected and not contained in $B(x, \frac{\varepsilon}{2}r)$, a point $x_2 \in A_n$ such that

$$d(x, x_2) = \frac{\varepsilon}{2} r.$$

Let y be the middle point between x_1 and x_2, and consider the ball B centered at y whose boundary contains x_1 and x_2. If x_1 and x_2 are connected by $A_n \cap B$, then we take $B = B_n$ and (10.11) is obviously satisfied (with $c = \frac{\varepsilon}{8}$), since then

$$(10.21) \qquad \mathcal{D}^1_{A_n}(B_n) \geq 1 \,.$$

(Indeed, by Relation 6.32 and the remarks following it, we have in this case $\mathcal{H}^1(A_n \cap B_n) \geq d(x_1, x_2)$.) If this is not the case, we let the radius of B grow until x_1 and x_2 become connected by $A_n \cap B$ or B touches the boundary of $\hat{B}$ and we chose such a ball as B_n. In the first case x_1 and x_2 are both connected to a point x_3 on ∂B_n. Then

$$\mathcal{H}^1(A_n \cap B) \geq d(x_1, x_3) + d(x_2, x_3) \,.$$

Since x_1 and x_2 are symmetric with respect to y, the above inequality again leads to (10.21). In the last case, x_1 and x_2 are separately connected to ∂B_n by A_n.

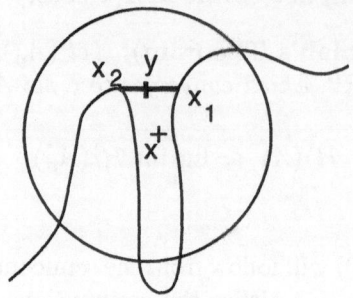

Figure 10.1

So we have

$$\mathcal{H}^1(A_n \cap B_n) \geq d(x_1, \partial B_n) + d(x_2, \partial B_n) \geq \text{diam } B_n - \frac{\varepsilon r}{2}.$$

Since the radius of B_n is clearly larger or equal to $(1 - \frac{\varepsilon}{2})r$, we get

$$\mathcal{H}^\alpha(A_n \cap B_n) \geq (1 - \frac{\varepsilon}{2(1 - \frac{\varepsilon}{2})})\text{diam}\,B_n \geq (1 - \varepsilon)\text{diam}\,B_n.$$

So we again obtain (10.11) with $c \geq \frac{1}{2}$.   □

Chapter 11

RECTIFIABLE SETS

In this chapter, we introduce the notion of rectifiable sets and mainly show that rectifiable sets are regular. The converse implication will be proved in the next chapter. We then discuss with some more details the case of rectifiable curves, which was already considered in Chapter 5.

1. Rectifiable surfaces and rectifiable sets.

In this chapter, we shall always assume that α is an integer. We call α-*rectifiable surface* any subset of $\mathbb{R}^N$ which is equal to the range of a Lipschitz map defined on a measurable subset of $\mathbb{R}^\alpha$.

An α-set A will be said α-*rectifiable* if it can be almost covered by a sequence of α-rectifiable surfaces. On the contrary, an α-set A will be said to be *purely unrectifiable* if it has a zero α-measure intersection with any α-rectifiable surface. Note that since a Lipschitz map defined on a subset of $\mathbb{R}^\alpha$ can be extended to all of $\mathbb{R}^\alpha$, we could have defined rectifiable surfaces as ranges of Lipschitz maps defined on $\mathbb{R}^\alpha$ and the above two definitions would keep the same meaning.

We can immediately deduce from the definitions that an α-set A cannot be at the same time rectifiable and purely unrectifiable, because otherwise it would have zero measure. It is also clear that a subset of a rectifiable set is rectifiable, a subset of a purely unrectifiable set is purely unrectifiable, the union of a sequence of rectifiable sets is rectifiable, and the union of a sequence of purely unrectifiable sets is purely unrectifiable.

Lemma 11.1. *An α-set A can be decomposed into the union of a rectifiable set A_1 and a purely unrectifiable set A_2.*

Proof. Let c be the maximum measure of a rectifiable subset of A. Since A is measurable, we can find a rectifiable subset A' of A with $\mathcal{H}^\alpha(A') = c$, by taking the union of the terms of a sequence $(A_n)_{n\in\mathbb{N}}$ such that $\mathcal{H}^\alpha(A_n) \to c$. Then $A_2 = A \setminus A'$ must be purely unrectifiable. Indeed, if S were a rectifiable surface such that $\mathcal{H}^\alpha(S \cap A_2) > 0$, $A_1 \cup (S \cap A_2)$ would be a rectifiable part of A with a measure larger than c. $\quad\square$

Let us state a simple rectifiability criterion.

Lemma 11.2. *Assume that A is an α-set and that W is a subspace of $\mathbb{R}^N$ with dimension $N - \alpha$ such that, for every $(x,y) \in A \times A$, $x \neq y$, $y - x$ makes an angle with W larger than a fixed value $\varphi > 0$. Then A is rectifiable.*

Proof. Let P be the orthogonal projector from A on $W^\perp$. One easily sees that P has an inverse map on $P(A)$, with a Lipschitz constant bounded by $(\sin \varphi)^{-1}$. $\quad \square$

2. Regularity of rectifiable sets.

We observe that rectifiable sets have been defined for integral dimensions, which are the ones for which nonnegligible regular sets appear. This chapter and the beginning of the next one are devoted to the study of the connection between the definition of regularity and that of rectifiability. Let us begin by examining a very particular case.

Lemma 11.3. *The range of a Lipschitz injective map from $\mathbb{R}$ to $\mathbb{R}^N$ is 1-regular.*

Proof. Fix t in $\mathbb{R}$ and let $r > 0$ be such that $f(t-1)$, $f(t+1) \notin B_r(f(t))$. Let t_1 be the maximum value of $s < t$ and t_2 the minimum value of $s > t$ such that $f(s) \notin B_r(f(t))$. Then $f(t_1), f(t_2) \in \partial B_r(f(t))$ and so, using Relation 6.32,

$$\mathcal{H}^1(f(\mathbb{R}) \cap B_r(f(t))) \geq \mathcal{H}^1(f([t_1,t])) + \mathcal{H}^1(f([t,t_2])) \geq$$
$$d(f(t_1),f(t)) + d(f(t),f(t_2)) = 2r = \operatorname{diam}(B_r(f(t))) .$$

(Indeed, by relation 6.32,

$$2r = \mathcal{H}^1([f(t_1),f(t_2)]) \leq \mathcal{H}^1(f([t_1,t_2])). \quad)$$

Therefore, for r small enough, $\mathcal{D}_{f(\mathbb{R})}(B_r(f(t))) \geq 1$. $\quad \square$

Let α be a fixed integer and let $f : \mathbb{R}^\alpha \to \mathbb{R}^N$ be a Lipschitz map. Consider the map $f^* : \mathbb{R}^\alpha \to \mathbb{R}^{N+\alpha}$ defined by $f^*(x) = (x, f(x))$.

Lemma 11.4. *A Lipschitz map f is differentiable at a point $x \in \mathbb{R}^\alpha$ if and only if the graph of f, namely the set $f^*(\mathbb{R}^\alpha)$, has a tangent space at $f^*(x)$ in the sense of (9.70).*

Proof. Let us begin by some preliminary considerations. If V is an α-dimensional space tangent to the graph of f, then V has to be the graph of an affine map g with a Lipschitz constant smaller or equal to the Lipschitz constant M of f. Then, the fact that a point $f^*(y)$ belongs to a certain angular neighbourhood C of V is equivalent to the condition

$$(11.5) \qquad |f(y) - g(y)| < \delta |y - x|$$

in correspondence with a suitable value of δ. A second remark is following: Assume that some point $z \in \mathbb{R}^\alpha$ satisfies

$$(11.6) \qquad |f(z) - g(z)| > 3\delta |z - x| .$$

Then (11.5) cannot be true for any point y in the ball B centered at z and with radius $\frac{\delta}{2M} |z - x|$, provided $\delta \leq M$. Indeed,

$$|f(y) - g(y)| \geq |f(z) - g(z)| - |f(z) - f(y)| - |g(z) - g(y)|$$
$$\geq 3\delta |z - x| - 2M d(y,z) > \delta |z - x|.$$

This means that

$$(11.7) \qquad f^*(B) \cap C = \emptyset .$$

Let us now begin with the proper proof. Assume first that $f^*(\mathbb{R}^\alpha)$ has a tangent space V at $f^*(x)$ and let us prove that g is the differential of f at x. If, in correspondence with a given value of δ, we can find a point z, as close to x as we want, such that (11.6) holds, then we also have (11.7). If P is the orthogonal projection from $\mathbb{R}^{N+\alpha}$ onto $\mathbb{R}^\alpha \times \{0\}$, we have by relation (6.32)

$$(11.8) \quad \mathcal{H}^\alpha(f^*(B)) \geq \mathcal{H}^\alpha(P(f^*(B))) = \mathcal{H}^\alpha(B) = c_\alpha (\frac{\delta}{M} |z - x|)^\alpha .$$

On the other hand, B is contained in the ball B' of $\mathbb{R}^\alpha$ centered at x with radius $2|z - x|$. Therefore $f^*(B)$ is contained in the ball B'' of $\mathbb{R}^{N+\alpha}$, centered at $f^*(x)$ and with radius $2(1+M)|z-x|$. From (11.7) and (11.8) we obtain

$$(11.9) \quad \mathcal{H}^\alpha(f^*(\mathbb{R}^\alpha) \cap B'' \setminus C) \geq \mathcal{H}^\alpha(f^*(B)) \geq c_\alpha \left(\frac{\delta \operatorname{diam} B''}{2M(1+M)} \right)^\alpha .$$

By (9.70) we see that (11.9) cannot hold, whatever C is, if the diameter of B'' is small enough, and therefore (11.6) cannot hold whatever δ is for z close enough to x. Thus, g is the differential of f at x.

The converse implication is achieved by observing that if g is the differential of f then, for every δ, (11.5) must hold for y close to x. That means that $f^*(\mathbb{R}^\alpha) \subset C$ on a neighbourhood of $f^*(x)$. $\quad\square$

Let us point out a variant of the main argument in the above proof. We have shown how (11.6) leads to a contradiction to the tangency assumption on f^* and V. However, the full tangency is not essential for the argument. In fact, assume that l is a straight line which makes a small angle with $z - x$. In other words, assume that the trace of B on l is a segment with a length of the same order of magnitude as the diameter of B. Then the same contradiction (11.9) can be derived from the tangency between the traces of V and $f^*(\mathbb{R}^\alpha)$ on $l \times \mathbb{R}^N$. Moreover the above proof also shows that this is going to happen when the restriction of f on l is differentiable and its differential is equal to the restriction of the linear affine map which has V as a graph. This fact has an important consequence. It shows that if we find a linear continuous map L such that its restriction to every straight line in a dense set of straight-lines crossing x is the differential at x of the restriction of f, then L is the differential of f at x. By combining such a consideration with the above lemmas, we can deduce the following statement.

Lemma 11.10. *A Lipschitz map from $\mathbb{R}^\alpha$ to $\mathbb{R}^N$ is differentiable at almost every $x \in \mathbb{R}^\alpha$.*

Proof. If $\alpha = 1$, the statement follows from Lemmas 11.3 and 11.4 and Proposition 9.68. In the general case, we fix a unitary vector j in $\mathbb{R}^\alpha$, and we consider the set A of the points of $\mathbb{R}^\alpha$ at which f does not have a derivative in the direction of j. We can easily see that A is measurable and the case $\alpha = 1$ shows that the trace of A has a one-dimensional measure zero on every straight line with the direction of j.

So, by Fubini's Theorem, we can conclude that A is a negligible subset of $\mathbb{R}^\alpha$. Moreover, for almost every x, the differential of f at x along the direction of j is given by the restriction of the distributional derivative at x. (This is not true in general, but in the case of a Lipschitz map it is easily fixed.) We can repeat the argument by letting j vary in a dense subset of the unit sphere of $\mathbb{R}^\alpha$. Then the discussion following Lemma 11.4 shows that the distributional derivative of f gives the differential at almost every x of $\mathbb{R}^\alpha$. $\quad\square$

Lemma 11.11 (Sard Lemma). *Let f be a Lipschitz map from $\mathbb{R}^\alpha$ to $\mathbb{R}^N$. If A' denotes the set of the points $x \in \mathbb{R}^\alpha$ at which f is not differentiable or the differential of f is not injective, then*

$$\mathcal{H}^\alpha(f^*(A')) = 0.$$

Proof. Let A'' be equal to the set of the points of A where f is differentiable but the differential is not injective. Fix $\varepsilon > 0$ and x in A''. If we take a small ball B around x, by the definition of differentiability we see that, if r denotes the radius of B and if M is the Lipschitz constant of f, then $f(B)$ is contained in the $\varepsilon' r$-neighbourhood of an $(\alpha - 1)$-dimensional ball with radius Mr, provided r is sufficiently small in dependence of ε'. Then, for a suitable choice of ε' we can find a covering $\mathcal{A}_B$ of $f(B)$ such that $\rho(\mathcal{A}_B) < \overline{\rho}$, for a suitable choice of the constant $\overline{\rho}$ and

$$(11.12) \qquad |\mathcal{A}_B|_\alpha < \varepsilon \mathcal{H}^\alpha(B).$$

So we can find a Vitali covering $\mathcal{B}$ of A'' consisting of closed balls B which satisfy (11.12). Fix a real number $R > 0$. By Vitali Covering Lemma 7.14 we can find a subset $\mathcal{B}'$ of $\mathcal{B}$, consisting of disjoint sets contained in $B_R(0)$, which is an almost covering of $A'' \cap B_R(0)$. Since, by the previous lemma, $\mathcal{H}^\alpha(A' \setminus A'') = 0$, then $\mathcal{B}'$ also is an almost covering of $A' \cap B_R(0)$. So we have by (11.12) and by Lemma 7.4, provided the value of $\overline{\rho}$ is suitably small,

$$\mathcal{H}^\alpha(f(A' \cap B_R(0))) = \mathcal{H}^\alpha(f(A' \cap B_R(0) \setminus \bigcup \mathcal{B}')) + \mathcal{H}^\alpha(f(\bigcup \mathcal{B}'))$$

$$= \mathcal{H}^\alpha(f(\bigcup \mathcal{B}')) \leq \sum_{B \in \mathcal{B}'} \mathcal{H}^\alpha(f(B)) \leq$$

$$\leq \sum_{B \in \mathcal{B}'} |\mathcal{A}_B|_\alpha + \varepsilon \leq \varepsilon \sum_{B \in \mathcal{B}'} \mathcal{H}^\alpha(B) + \varepsilon$$

$$\leq \varepsilon \mathcal{H}^\alpha(B_R(0)) = \varepsilon 2^\alpha c_\alpha R^\alpha.$$

By the arbitrariness of ε and R, we obtain the thesis from the above estimate. $\quad\square$

Lemma 11.11 reduces the problem of proving the regularity of $f(\mathbb{R}^\alpha)$ to checking the density of the image at a point where f has an injective differential.

Lemma 11.13. *Let f be a Lipschitz map from $\mathbb{R}^\alpha$ into $\mathbb{R}^N$, and let x be a point of $\mathbb{R}^\alpha$ at which f is differentiable and such that the differential of f at x is an injection. Then*

$$(11.14) \qquad \underline{d}^\alpha_{f(\mathbb{R}^\alpha)}(f(x)) \geq 1 .$$

Proof. Let V be the range of the differential of f at x. By a linear change of variable we can define f on V and assume that the differential at x is given by the restriction on V of the identity map of $\mathbb{R}^N$. We can also assume without any restriction that $x = 0$ and $f(x) = 0$. By the definition of differentiability we see that for any given $\varepsilon > 0$ we have for r small enough,

$$(11.15) \qquad \forall\, y \in V \cap B_r(0) : \quad |f(y) - y| \leq \varepsilon|y| .$$

This fact implies, by the Brouwer Fixed Point Theorem, that

$$(11.16) \qquad V \cap B_{(1-\varepsilon)r}(0) \subset P_V(f(V \cap B_r(0))) .$$

Indeed, if $z \in V \cap B_{(1-\varepsilon)r}(0)$, the map defined on $V \cap B_r(0)$ by

$$F(y) \,=\, y - P_V(f(y)) + z$$

sends $V \cap B_r(0)$ in itself as a consequence of (11.15). The Brouwer theorem states that any continuous map from a ball into itself has a fixed point. So we can find a fixed point y for F and $F(y) = y$ means $P_V(f(y)) = z$. Moreover, from (11.15) we also deduce that

$$(11.17) \qquad f(V \cap B_r(0)) \subset B_{(1+\varepsilon)r}(0) .$$

From (11.16) and (11.17) we get

$$\begin{aligned} \mathcal{H}^\alpha(f(\mathbb{R}^\alpha) \cap B_{(1+\varepsilon)r}(0)) &\geq \mathcal{H}^\alpha(f(V \cap B_r(0))) \\ &\geq \mathcal{H}^\alpha(P_V(f(V \cap B_r(0)))) \geq \mathcal{H}^\alpha(V \cap B_{(1-\varepsilon)r}(0)) . \end{aligned}$$

By the arbitrariness of r we deduce that

$$\underline{d}^\alpha_{f(\mathbb{R}^\alpha)}(0) \,\geq\, (\frac{1+\varepsilon}{1-\varepsilon})^\alpha .$$

We can pass to the limit with respect to ε and, since we have assumed $f(x) = 0$, we obtain (11.14).   □

We can combine the last two lemmas in order to prove the regularity of the rectifiable sets.

Theorem 11.18. *If f is a Lipschitz map from $\mathbb{R}^\alpha$ to $\mathbb{R}^N$, then $f(\mathbb{R}^\alpha)$ is an α-regular subset of $\mathbb{R}^N$.*

From the above statement and from Corollary 8.25 we conclude that a rectifiable set cannot contain a fully unregular subset with positive measure, and therefore we also obtain the main result of this chapter.

Corollary 11.19. *Every α-rectifiable set is α-regular.*

3. Rectifiable curves.

In order to connect the particular case of rectifiable sets of dimension 1 with the notion of rectifiable curves already used in Chapter 5, we consider a minor variant of the definitions in the beginning of this chapter. We shall call α-*simply rectifiable surface* any subset of $\mathbb{R}^N$ which is equal to the range of an injective Lipschitz map defined on a measurable subset of $\mathbb{R}^\alpha$. Analogously, an α-set A will be called an α-*simply rectifiable set* if it can be almost covered by a sequence of α-simply rectifiable surfaces and *purely α-unrectifiable* if it has a zero measure intersection with any simply rectifiable surface. When $\alpha = 1$ we shall speak of *simple curves* in order to distinguish 1-simply rectifiable surfaces which are ranges of injective Lipschitz maps defined on an interval. Of course we can assume after a rescaling that such an interval has length 1. In such a case, we shall call *length* of a curve c the least Lipschitz constant of all the Lipschitz functions defined on an interval with length 1 whose range is equal to c. An alternative definition which leads to the same concept is to define the length of a curve as the smallest length of an interval on which we can define a map with Lipschitz constant 1 whose range is equal to the curve. We shall denote by $l(c)$ the length of a curve c.

All this has been detailed in Chapter 5, Section 2. Let us now look at the relation between l and $\mathcal{H}^1$. By (6.32) we immediately see that $\mathcal{H}^1(c) \leq l(c)$. Let us show that in the case of a simple curve the reverse inequality also holds. We may assume that our curve is equal to the range of a one-to-one map defined on a closed interval; the general case can then be covered by passing to a countable union. So we can transfer to c the natural ordering of the closed interval. Let ε be a given positive constant, and let $\mathcal{A} \in \mathcal{C}(c)$ be such that $|\mathcal{A}| \leq \mathcal{H}^1(c) + \varepsilon$ and that $\rho(\mathcal{A}) \leq \varepsilon$. We can assume that all the sets in $\mathcal{A}$ are open and therefore, by compactness, we may assume that $\mathcal{A}$ is a finite set. We can find $A_1 \in \mathcal{A}$ such that the minimum point x_1 of c belongs to A_1. Let x_2 be the upper bound (for the ordering of c) of the points of c which belong to A_1. Then we fix a set $A_2 \in \mathcal{A}$ such that $x_2 \in A_2$, and we call x_3 the upper bound of the points of c which are in A_2.

We can proceed recursively in this way and, after a finite number of steps, reach a point x_n which necessarily is the last for the ordering of c. We then define a piecewise linear function f_ε on the interval $[0, \sum_{i=1}^{n-1} \text{diam}(A_i)]$ by interpolating the points x_j and x_{j+1} on the interval $[\sum_{i=1}^{j-1} \text{diam}(A_i) \sum_{i=1}^{j} \text{diam}(A_i)]$. Such a function clearly has a Lipschitz constant less or equal to 1. Moreover, its range is contained in an ε-neighbourhood of c. We can reparametrize the function f_ε on the interval $[0, 1]$ and we obtain a Lipschitz constant less or equal than $\sum_{i=1}^{n-1} diam(A_i) \leq |\mathcal{A}| \leq \mathcal{H}^1(c) + \varepsilon$. Then, we let ε tend to zero and we take a uniform limit function by using the Ascoli-Arzela Theorem from Chapter 5.2 (provided $\mathcal{H}^1(c) < \infty$, otherwise we have nothing to prove). We obtain in this way a function with a Lipschitz constant less or equal to $\mathcal{H}^1(c)$ whose range is contained in c. Moreover, since the range of such a function contains the extremal points of c and c is simple, the range has to be actually equal to c. The existence of such a function yields the reverse inequality

$$l(c) \leq \mathcal{H}^1(c),$$

and therefore the equality of both measures on simple curves.

This last result will imply a useful property of α-simply rectifiable sets: they are almost equal to a countable union of simple curves. The precise statement is given in Theorem 11.23 below. Let us consider a Borel subset A of a simple curve c and let f be a one-to-one Lipschitz map defined on $[0, 1]$ and with a Lipschitz constant equal to the length of c and whose range is equal to c. Then $f^{-1}(A)$ is a Borel subset of $[0, 1]$ and therefore, for every positive constant ε, it can be covered by a sequence of intervals the sum of whose lengths is smaller than its 1-dimensional Lebesgue measure plus ε. We denote by $(c_i)_{i \in \mathbb{N}}$ the sequence of the curves obtained by taking the images by f of such intervals. Since the Lipschitz constant of f is equal to $l(c)$, we obtain

$$(11.20) \qquad \mathcal{H}^1(A) \ \leq \ \sum_i l(c_i) \ \leq \ l(c)(\mathcal{L}_1(f^{-1}(A)) + \varepsilon).$$

The above inequality implies in particular, if we let ε tend to zero, that

$$(11.21) \qquad \mathcal{H}^1(A) \ \leq \ l(c)\mathcal{L}_1(f^{-1}(A)).$$

Inequality (11.21) can be clearly reversed, since the reverse inequality is trivially equivalent to (11.21) applied to the set $c \setminus A$. If we plug such an estimate into (11.20) we find

$$(11.22) \qquad \sum_i l(c_i) \ \leq \ \mathcal{H}^1(A) + \varepsilon l(c).$$

By the results of Chapter 6, any simply rectifiable 1-set A can be decomposed into the union of a $\mathcal{H}^1$-negligible set A' and of a Borel set A'' which can be completely covered by a sequence of simple curves. Of course this allows us to write A'' as a countable union of disjoint Borel subsets A_i of simple curves. Given $\varepsilon \geq 0$, for each one of the indices i, we can apply (11.22) in order to cover it with a sequence of curves such that the sum of their lengths is smaller than or equal to $\mathcal{H}^1(A_i) + \frac{\varepsilon}{2^i}$. We can take all these curves, considered for every value of i, as terms of a sequence $(c_j)_{j \in \mathbb{N}}$. In this way we get the following useful result.

Theorem 11.23. *If A is a simply rectifiable 1-set, we can find an almost covering $(c_j)_{j \in \mathbb{N}}$ of A, consisting of simple curves, such that*

$$(11.24) \qquad \sum_i l(c_i) \leq \mathcal{H}^1(A) + \varepsilon.$$

Chapter 12

PROPERTIES OF REGULAR
AND RECTIFIABLE SETS

The full equivalence between the notions of rectifiability and regularity is developed in this chapter, as well as two very remarkable geometric properties of unrectifiable sets. It is first proved that there is a universal constant $c(\alpha)$ such that at almost every point of a fully unrectifiable set, the lower conic density is larger that $c(\alpha)$. This is a local and clear cut computational criterion for distinguishing rectifiable points from unrectifiable points in a given α-set. From this property, and from the existence almost everywhere of a tangent space for regular sets, it is immediately deduced that fully unrectifiable sets are fully irregular and that regular sets are rectifiable. We also know from Chapter 11 that rectifiable sets are regular. Thus the equivalence between regularity and rectifiability will be complete and, as a by product, the equivalence of simple rectifiability and rectifiability. We end the chapter with the surprising property of $(N-1)$-fully irregular (or unrectifiable) sets of $\mathbb{R}^N$ to have a negligible projection on almost every hyperplane.

1. Geometric lemmas.

Let W be a set of vectors of $\mathbb{R}^N$, R and φ be real positive numbers, $\varphi < \frac{\pi}{2}$ and let $x \in \mathbb{R}^N$. We shall denote by $C(x, W, \varphi)$ the set of the points y in $\mathbb{R}^N$ such that $y - x$ makes an angle smaller than φ with some vector in W. We denote by $C(x, W, \varphi, r)$ the set $C(x, W, \varphi) \cap B_r(x)$.

Let us fix W as a linear subspace of $\mathbb{R}^N$ and $\varphi > 0$. For x and y in $\mathbb{R}^N$, we denote by $C(x, y)$ the set $C(x, W, \varphi) \cup C(y, W, \varphi)$. We also fix a positive number $\psi < \varphi$ and let $\mathcal{R}$ be the set of the pairs $(x, y) \in \mathbb{R}^N \times \mathbb{R}^N$ such that $y - x$ makes an angle smaller than ψ with W, namely such that $y \in C(x, W, \psi)$ or equivalently $x \in C(y, W, \psi)$.

Set $\mu = \frac{1}{6}(\sin \varphi - \sin \psi)$. We begin with a simple geometric lemma (see Figure 12.1).

Lemma 12.1. *Let $(x, y) \in \mathcal{R}$ and $d(x, y) = R$. Then*

(12.2)
$$B_{3\mu R}(\frac{1}{2}x + \frac{1}{2}y) + W \subset C(x, y).$$

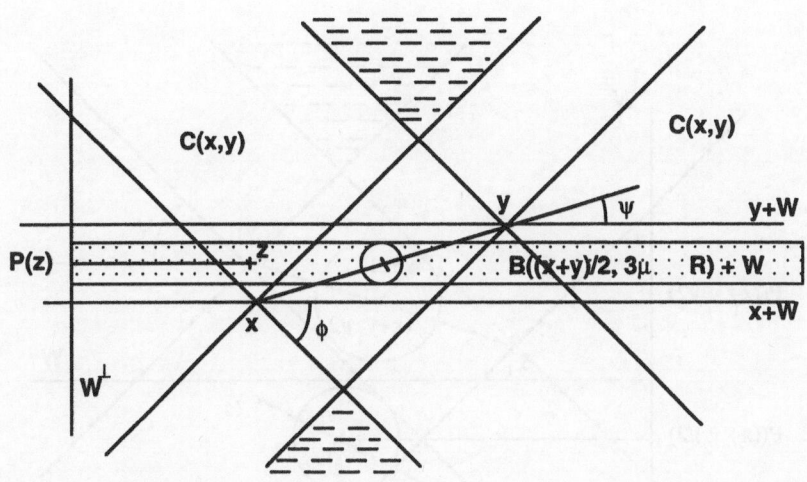

Figure 12.1

Proof. Fix z in $B_{3\mu R}(\frac{1}{2}x + \frac{1}{2}y) + W$. Then the distance between the projections on $W^{\perp}$ of z and of the middle point of x and y is bounded by $3\mu R$. The projection of the middle point $\frac{1}{2}x + \frac{1}{2}y$ and the projection of x on $W^{\perp}$ have a distance smaller than $\frac{1}{2}R \sin \psi$, since $d(x,y) = R$ and $(x,y) \in \mathcal{R}$. So we deduce that the distance between the projections $P(x)$ and $P(z)$ on $W^{\perp}$ of x and z is bounded by $\frac{1}{2}R \sin \psi + 3\mu R = \frac{1}{2}R \sin \varphi$. Since we can assume (by exchanging, if necessary, the roles of x and y) that $d(x,z) \geq \frac{1}{2}R$, we obtain $\frac{d(P(x),P(z))}{d(x,z)} \leq \sin \varphi$. Thus $z \in C(x, W, \varphi)$ and therefore $z \in C(x, y)$. $\square$

For $(x, y) \in \mathcal{R}$ and $R = d(x, y)$ we set

$$B(x, y) = B_{\mu R}(\frac{1}{2}x + \frac{1}{2}y) + W .$$

An immediate consequence of the previous lemma is the following statement, provided we assume that ψ is taken small enough to ensure $\sin \psi < 2\mu$.

Lemma 12.3. *If (x_1, y_1) and $(x_2, y_2) \in \mathcal{R}$, $d(x_2, y_2) \leq d(x_1, y_1)$ and $x_2 \notin C(x_1, y_1)$, then $B(x_1, y_1) \cap B(x_2, y_2) = \emptyset$.*

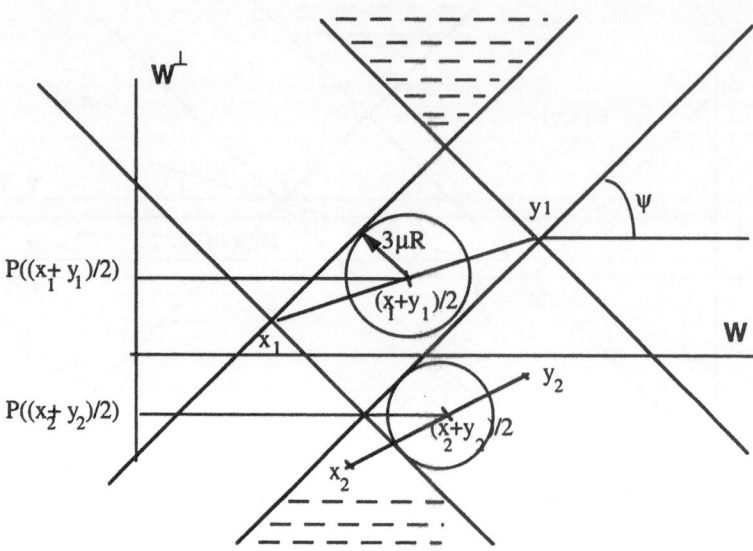

Figure 12.2 (Lemma 12.3)

Proof. Set $R = d(x_1, y_1)$. Since $x_2 \notin C(x_1, y_1)$ by the previous lemma, the projection $P(x_2)$ of x_2 on $W^\perp$ must have a distance from the projection of $\frac{1}{2}x_1 + \frac{1}{2}y_1$ larger than or equal to $3\mu R$. Since $d(x_2, \frac{1}{2}x_2 + \frac{1}{2}y_2) = \frac{1}{2}d(x_2, y_2) \le \frac{1}{2}R$, the distance between the projections of x_2 and $\frac{1}{2}x_2 + \frac{1}{2}y_2$ on $W^\perp$ is bounded by $\frac{1}{2}R \sin \psi$ and therefore by μR. (Recall that we fixed $\sin \psi < 2\mu$.) So the distance between the projections on $W^\perp$ of the two middle points $\frac{1}{2}x_2 + \frac{1}{2}y_2$ and $\frac{1}{2}x_1 + \frac{1}{2}y_1$ is larger than $2\mu R$, because $d(x_1, y_1) \le R$, $d(x_2, y_2) \le R$. The thesis follows.   □

2. Equivalence of the notions of regularity and rectifiability.

The remarkable geometric argument which leads to the result of Theorem 12.4 is essentially due to Besicovitch. In order to capture its main idea, let us return to the meaning of Lemma 11.2. Acccording to this lemma, if a set A is such that any pair x, y in it defines a line making an angle less than $0 < \varphi < \frac{\pi}{2}$ with some fixed direction W, then A is a Lipschitz graph and therefore simply rectifiable. We can state this result by using cones $C(W, x, \varphi, r)$ whose vertex x is on A. If there is a direction W and a part F of A with positive measure such that for any x in F and for some r, the cone $C(x, W, \varphi, r)$ does not meet A, then this part must be simply rectifiable by Lemma 11.2 and A cannot be fully simply unrectifiable. From this "soft" argument, we can deduce that at almost every point x of an unrectifiable set A, the cones $C(x, W, \varphi, r)$ meet A for every φ and r. Now, as shows the next theorem, *this argument can be made quantitative* and yields a universal lower estimate for the angular density of an unrectifiable set at almost every point.

Assume that α is an integer, that A is an α-set and that W is a subspace of $\mathbb{R}^N$ of dimension $N - \alpha$.

Theorem 12.4. *There is a constant $c > 0$ (e.g., $c = \frac{c_\alpha}{4.12^\alpha}$), only depending on α, such that, for every fully simply unrectifiable set A, for every $\varphi > 0$ and for every subspace W of dimension $N - \alpha$, the upper conic density*

$$d(x, W, \varphi) = \limsup_{R \to 0} \frac{\mathcal{H}^\alpha(A \cap C(x, W, \varphi, R))}{(R \sin \varphi)^\alpha}$$

is larger than c at almost every x in A.

Proof. Assume that c is a positive constant which does not satisfy the thesis. Then, for some $\varphi > 0$, we can choose a $(N - \alpha)$-dimensional subspace W and a positive number $\overline{R}$ in such a way that the set F' of the points x of A such that

(12.5) $\qquad \forall R < \overline{R} : \quad \mathcal{H}^\alpha(A \cap C(x, W, \varphi, R)) < c(R \sin \varphi)^\alpha$

has a positive measure. By Proposition 6.26, we can find a closed subset F of F' which still has a positive measure and therefore by (8.18) a ball B with radius $\overline{R}$ (by eventually replacing $\overline{R}$ by a smaller value) such that

(12.6) $\qquad \mathcal{D}_F(B) \geq 2^{-(\alpha+1)} .$

We now fix a convenient value of $\psi < \varphi$ and we consider the set $\mathcal{R}$ of pairs (x, y) defined at the beginning of this chapter.
We denote by (x_1, y_1) a pair in $\mathcal{R} \cap ((F \cap \overline{B}) \times (F \cap \overline{B}))$ which maximizes the value of $R_1 = d(x_1, y_1)$, provided $\mathcal{R} \cap ((F \cap \overline{B}) \times (F \cap \overline{B}))$ is not empty. Then we define recursively a (possibly finite) sequence of pairs $(x_1, y_1), (x_2, y_2), \ldots, (x_n, y_n), \ldots$ by choosing a pair (x_n, y_n) in $\mathcal{R} \cap ((F \cap B) \setminus \bigcup_{i=1}^{n-1} C(x_i, y_i))^2$ (if this set is not empty) such that the value of $R_n = d(x_n, y_n)$ is maximum.

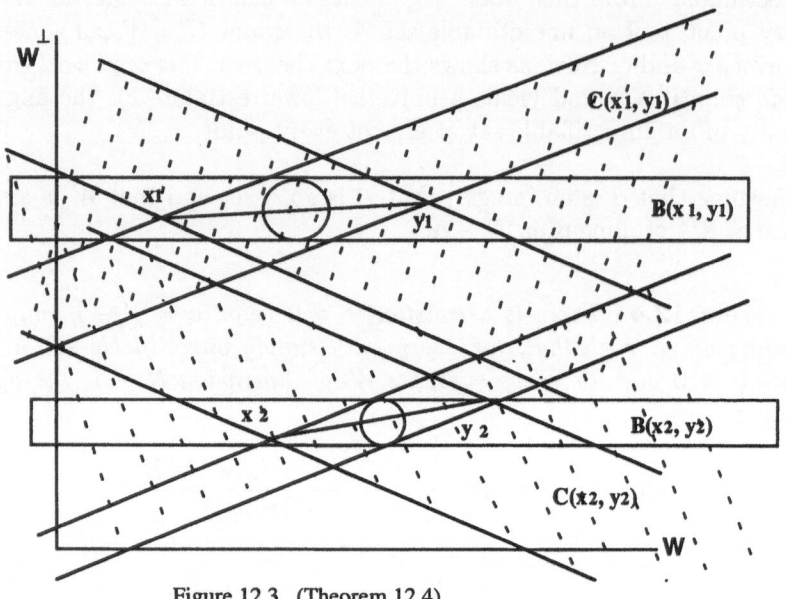

Figure 12.3 (Theorem 12.4)

By Lemma 12.3, the sets $B(x_n, y_n)$ defined above turn out to be disjoint and by their definition this fact implies that they have disjoint projections on $W^{\perp}$. The projection of $B(x_n, y_n)$ on $W^{\perp}$ is a ball with radius μR_n and it is contained, by construction, in the ball B' whose center is the projection on $W^{\perp}$ of the center of B and whose radius is $(1 + \mu)\overline{R}$. Therefore we have

$$(12.7) \qquad\qquad \sum_n (\mu R_n)^{\alpha} \leq (1 + \mu)^{\alpha} \overline{R}^{\alpha}.$$

Note that if R_n exists for every n, then $R_n \to 0$. By the maximality of $d(x_n, y_n) = R_n$ we have

$$B \cap F \cap C(x_n, W, \varphi) \setminus \bigcup_{i=1}^{n-1} C(x_i, y_i) \subset C(x_n, W, \varphi, R_n) \,.$$

Thus, returning to the definition of the $C(x_i, y_i)$,

$$B \cap F \cap C(x_n, W, \varphi) \setminus \bigcup_{i=1}^{n-1} (C(x_i, W, \varphi) \cup C(y_i, W, \varphi)) \subset C(x_n, W, \varphi, R_n) \,.$$

Using the obvious identity $B \cap \bigcup_n C_n = \bigcup_n (B \cap (C_n \setminus \bigcup_{i=1}^{n-1} C_i))$, we obtain

$$B \cap F \cap \bigcup_n C(x_n, y_n) \;=\; B \cap F \cap \bigcup_n (C(x_n, W, \varphi) \cup C(y_n, W, \varphi))$$

$$= B \cap F \cap \bigcup_n (C(x_n, W, \varphi, R_n) \cup C(y_n, W, \varphi, R_n)) .$$

By (12.5), since $x_n \in F'$ and $F \subset A$,

$$\mathcal{H}^\alpha (F \cap C(x_n, W, \varphi, R_n)) \;\leq\; \mathcal{H}^\alpha (A \cap C(x_n, W, \varphi, R_n)) \;\leq\; c(R_n \sin\varphi)^\alpha .$$

The same bound also holds with x_n replaced by y_n, so by adding all the terms we have

$$\mathcal{H}^\alpha \left(B \cap F \cap \bigcup_n C(x_n, y_n) \right) \;\leq$$

$$\leq \sum_n \mathcal{H}^\alpha (F \cap (C(x_n, W, \varphi, R_n) \cup C(y_n, W, \varphi, R_n)))$$

$$\leq 2c \sum_n (R_n \sin\varphi)^\alpha .$$

We can combine this last inequality with (12.7), by taking into account that we can choose ψ such that $\mu = \frac{1}{6}(\sin\varphi - \sin\psi) \geq \frac{1}{6+\varepsilon} \sin\varphi$, in order to obtain

$$(12.8) \qquad \mathcal{H}^\alpha \left(B \cap F \cap \bigcup_n C(x_n, y_n) \right) \;\leq\; 2c\,(6+\varepsilon)^\alpha\,(1+\mu)^\alpha\,\overline{R}^\alpha .$$

On the other hand, if we consider the set $A' = B \cap F \setminus \bigcup_n C(x_n, y_n)$, we have by construction that

$$\forall\,(x,y) \in (A' \times A') \cap \mathcal{R} : \qquad d(x,y) \;\leq\; \inf_n R_n \;=\; 0 .$$

Since $\dim W^\perp = \alpha$, we can apply to A' Lemma 11.2 and we deduce that $\mathcal{H}^\alpha(A') = 0$, because A is purely simply unrectifiable. So the left-hand side of (12.8) is simply given by $\mathcal{H}^\alpha(B \cap F)$, and this estimate, combined with (12.6) yields

$$\frac{1}{2}\,c_\alpha\,\overline{R}^\alpha \;\leq\; 2c\,(6+\varepsilon)^\alpha\,(1+\mu)^\alpha\,\overline{R}^\alpha .$$

Thus, we obtain the required bound on c. Indeed, since ε can be arbitrarily small, and $(1+\mu)$ is always smaller than 2, we get

$$c \geq \frac{c_\alpha}{4.12^\alpha}$$

as a lower bound for c. □

An immediate consequence of the previous lemma is the nonexistence of a tangent space in the sense of Lemma 9.68.

Lemma 12.9. *If A is a fully simply unrectifiable set, then, for almost every x in A, relation (9.70) (i.e., $\underline{d}^\alpha_{A\backslash C}(x) = 0$) is not satisfied, whatever the choice of V is, for $C = C(x, V, \varphi)$, $\varphi < \frac{\pi}{2}$.*

Proof. In fact $\mathbb{R}^N \setminus C$ contains the set $C(x, W, \frac{\pi}{2} - \varphi)$, where $W = V^\perp$ has dimension $N - \alpha$. So, since $\frac{\pi}{2} - \varphi > 0$, (9.70) is in contradiction with Lemma 12.4, for almost every value of x. □

This last result produces a proof of the equivalence between the regularity and rectifiability properties.

Theorem 12.10. *If A is an α-set, then*
(i) A is regular if and only if α is an integer and A is an α-rectifiable set.
(ii) A is regular if and only if α is an integer and A is an α-simply rectifiable set.

Proof. If α is an integer and A is an α-rectifiable α-set, we know from Corollary 11.19 that A is regular. If A is regular, α must be an integer, as stated in Lemma 9.31, and for almost every x in A there is a tangent space to A. So, by the above lemma A cannot be purely simply unrectifiable. Since the same conclusion holds for any subset of A with positive measure we conclude that A must be simply rectifiable. Finally, if A is simply rectifiable it also obviously is rectifiable. □

3. Projection properties of irregular sets.

In the last part of the chapter we shall be concerned with the case $\alpha = N - 1$, which is the only one involved in the applications to image processing discussed in this book. This restriction will make the exposition much simpler.

We denote by Σ the set of the unitary vectors of $\mathbb{R}^n$. If $\theta \in \Sigma$, P_θ denotes the orthogonal projector on $\theta^\perp$. Let A be a purely unrectifiable $(N-1)$-set. Consider the product set $A \times \Sigma$ endowed with the product measure $\mathcal{H}^{N-1} \times \mathcal{H}^{N-1}$.

We shall say that a pair (x, θ) in $A \times \Sigma$ is a *concentration pair of the first kind* if x is an accumulation point for the set $(x + \mathbb{R} \cdot \theta) \cap A$. The fact that (x, θ) is a concentration pair of the first kind will be equivalently expressed by saying that θ is a concentration direction of the first kind for x or that x is a concentration point of the first kind for θ. Of course, (x, θ) is a concentration pair if and only if $(x, -\theta)$ is.

Lemma 12.11. *Let A be a $(N-1)$-set, $A' \subset A$ and let θ be a vector in Σ which is a concentration direction of the first kind for almost every point of A'. Then*

(12.12)
$$\mathcal{H}^{N-1}(P_\theta(A')) = 0 .$$

Proof. Since a projection can only decrease the Hausdorff measure, it is clear that for almost every y in $P_\theta(A')$, the set $A \cap P_\theta^{-1}(y)$ is an infinite set. This fact will imply the thesis. Let h, k be two integers and $B_{h,k}$ be the subset of $P_\theta(A')$ consisting of the points y such that $A \cap P_\theta^{-1}(y)$ contains at least k points at a distance larger or equal than $\frac{1}{h}$ from each other. For every k, $(B_{h,k})_{h \in \mathbb{N}}$ is an increasing sequence which almost covers $P_\theta(A')$. We divide the straight-line $\mathbb{R} \cdot \theta$ into a sequence of disjoint segments S_m with length $\frac{1}{h}$. Let $A_m = A \cap (S_m + \theta^\perp)$. Then

$$\sum_m \mathcal{H}^{N-1}(A_m) = \mathcal{H}^{N-1}(A) .$$

Every y in $B_{h,k}$ belongs by construction to $P_\theta(A_m)$ for at least k values of m. If we denote by f_m the characteristic functions of the $P_\theta(A_m)$, we see that their sum f satisfies $f = \sum_m f_m \geq k$ on $B_{h,k}$. So we have, by taking the integrals with respect to $\mathcal{H}^{N-1}$, namely, by Chapter 7, Section 3, with respect to the Lebesgue measure of $\theta^\perp$,

$$\mathcal{H}^{N-1}(B_{h,k}) \leq k^{-1} \int f \leq k^{-1} \sum_m \int f_m$$

$$= k^{-1} \sum_m \mathcal{H}^{N-1}(P_\theta(A_m))$$

$$\leq k^{-1} \sum_m \mathcal{H}^{N-1}(A_m) = k^{-1} \mathcal{H}^{N-1}(A) .$$

We can pass to the limit with respect to h and we get by (6.21) (and the subsequent remark), for a fixed value of k,

$$\mathcal{H}^{N-1}(P_\theta(A')) = \mathcal{H}^{N-1}(\bigcup_h B_{h,k}) = \sup_h \mathcal{H}^{N-1}(B_{h,k})$$

$$\leq k^{-1}\mathcal{H}^{N-1}(A) .$$

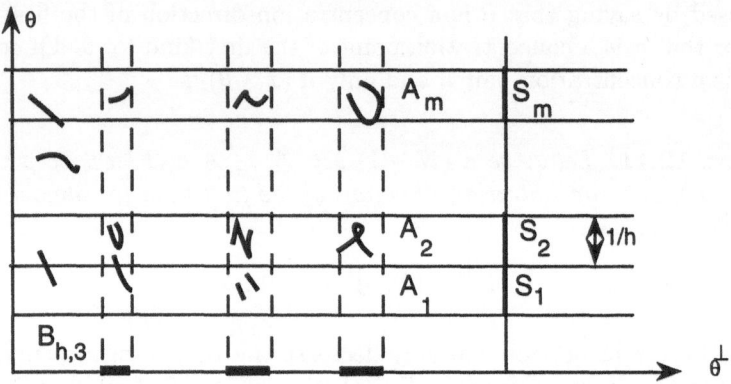

Figure 12.4 (Concentration direction of the first kind)

Since k is arbitrary, we obtain the thesis.   □

A pair (x, θ) in $A \times \Sigma$ will be called a *concentration pair of the second kind* if, for every constant $\delta > 0$, one can find a ball $\hat{B}$ around $P_\theta(x)$ in $\theta^\perp$ with arbitrarily small radius and a set X contained in $P_\theta^{-1}(\hat{B})$ such that

$$(12.13) \qquad \mathcal{H}^{N-1}(X \cap A) \geq \delta \mathcal{H}^{N-1}(\hat{B})$$

$$(12.14) \qquad \mathcal{H}^{N-1}(P_\theta(X)) \leq \delta^2 \mathcal{H}^{N-1}(\hat{B}) .$$

We can always assume, without any restriction, by using Lemma 6.26, that the set X is compact and, by eventually enlarging X, that (12.14) is an equality and therefore that the diameter of $P_\theta(X)$ is larger than constant multiplied by the diameter of $\hat{B}$. Indeed, since $P_\theta(X) \subset \mathbb{R}^{N-1}$, we have

$$c_{N-1}(2 \operatorname{diam}(P_\theta(X))^{N-1} \geq \mathcal{H}^{N-1}(P_\theta(X)) = \delta^2 c_{N-1}(\operatorname{diam}\hat{B})^{N-1}.$$

So, if we have a subset A' of A of concentration points of the second kind with respect to a given θ in Σ (namely a set A' such that every pair (x, θ) with x in A' is a concentration pair of the second kind), then the set $\mathcal{B}$ consisting of the sets $B = P_\theta(X)$, where X is obtained as above, contains closed sets and is a Vitali approximate covering of $P_\theta(A')$, according to Definition 7.13.

Combining (12.13) and (12.14) we obtain

$$(12.15) \qquad \forall B \in \mathcal{B} : \mathcal{H}^{N-1}(B) \leq \delta \mathcal{H}^{N-1}(A \cap P_\theta^{-1}(B))$$

$$(12.16) \quad \forall B \in \mathcal{B} : \quad \mathcal{H}^{N-1}(A \cap P_\theta^{-1}(B)) \geq \delta c_{N-1} (\text{diam } B)^{N-1} .$$

Lemma 12.17. *Let A be a $(N-1)$-set , $A' \subset A$ and let θ be a direction in Σ which is a concentration direction of the second kind for almost every point of A'. Then we again have (12.12), namely*

$$\mathcal{H}^{N-1}(P_\theta(A')) = 0 .$$

Proof. By Vitali Covering Lemma 7.14, since (12.16) gives a bound on $|\mathcal{B}'|_{N-1}$ for every collection $\mathcal{B}'$ of disjoint elements of $\mathcal{B}$, we can choose such a $\mathcal{B}'$ to be an almost covering of $P_\theta(A')$. So we have by (12.15),

$$\mathcal{H}^{N-1}(P_\theta(A')) \leq \sum_{B \in \mathcal{B}'} \mathcal{H}^{N-1}(B) \leq \delta \sum_{B \in \mathcal{B}'} \mathcal{H}^{N-1}(A \cap P_\theta^{-1}(B))$$

$$\leq \delta \mathcal{H}^{N-1}(A) .$$

Since δ can be taken arbitrary small, the above estimate yields (12.12).

$\square$

We say that a point x in A is a *radiation point* if almost every θ in Σ is a concentration direction (of the first or of the second kind) for x.

The next result is essentially a consequence of the angular lower density estimate obtained in Theorem 12.4.

Lemma 12.18. *If A is an unrectifiable $(N-1)$-set, then almost every point of A is a radiation point.*

Proof. By Theorem 12.4, we know that for almost every x in A and for all choices of W and φ, the lower angular density bound stated in Theorem 12.4 holds. Note that in our case the space W has dimension 1 and therefore W is determined by some θ in Σ. Using the same notation as in the beginning of the chapter, we set $C(x, \theta, \varphi, R) = C(x, W, \varphi, R)$. Of course, with this notation, $C(x, \theta, \varphi, R) = C(x, -\theta, \varphi, R)$. Let Σ_1 be the set of the elements of Σ which are concentration directions of the first kind for x. Fix θ in $\Sigma \setminus \Sigma_1$. By Corollary 8.17 we can assume that

$$(12.19) \qquad\qquad d_{\Sigma_1}^{N-1}(\theta) = 0 .$$

We are going to prove that θ must be a concentration direction of the second kind. Fix a constant $\delta > 0$. By (12.19) we can choose $\varphi > 0$ in such a way that the set Φ of the vectors in Σ_1 which make an angle with θ less than φ has a measure less than $\delta^2 c_{N-1}(2 \sin \varphi)^{N-1}$. If we denote by Φ_R, for $R > 0$, the set of the points $\frac{y-x}{|y-x|}$ obtained for $y \in A \cap C(x, \theta, \varphi, R)$, we have, by definition of Σ_1,

$$\Phi = \bigcap_{R > 0} \Phi_R .$$

This fact implies, for R small enough,

$$(12.20) \qquad\qquad \mathcal{H}^{N-1}(\Phi_R) \le \delta^2 c_{N-1}(2 \sin \varphi)^{N-1} .$$

Therefore

$$(12.21) \qquad \mathcal{H}^{N-1}(P_\theta(A \cap C(x, \theta, \varphi, R))) \le \delta^2 c_{N-1}(2R \sin \varphi)^{N-1} .$$

By Theorem 12.4 we can find arbitrarily small values of R for which

$$(12.22) \qquad \mathcal{H}^{N-1}(A \cap C(x, \theta, \varphi, R)) > c(R \sin \varphi)^{N-1} .$$

In order to obtain (12.13) and (12.14), we set

$$X = A \cap C(x, \theta, \varphi, R)$$
$$\hat{B} = P_\theta(C(x, \theta, \varphi, R)) .$$

So we have to prove that

$$(12.23) \qquad \mathcal{H}^{N-1}(A \cap C(x, \theta, \varphi, R)) \ge \delta \, \mathcal{H}^{N-1}(P_\theta(C(x, \theta, \varphi, R)))$$

$$(12.24) \qquad\qquad \mathcal{H}^{N-1}(P_\theta(A \cap C(x, \theta, \varphi, R))) \le \delta^2 \, \mathcal{H}^{N-1}(\hat{B}) .$$

Then (12.24) is equivalent to (12.21) while (12.23) follows from (12.22) for δ small enough with respect to the constant c of Theorem 12.4. Indeed, $\mathcal{H}^{N-1}(P_\theta(C(x, \theta, \varphi, R))) = c_{N-1}(2R \sin \varphi)^{N-1}$. Thus it is enough to take $\delta < \frac{c}{c_{N-1}2^{N-1}}$. $\square$

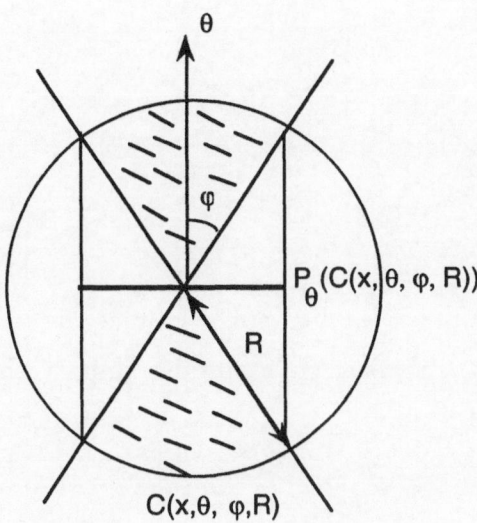

Figure 12.5 (illustrates Lemma 12.18)

We denote by R the set of the pairs (x, θ) in $A \times \Sigma$ which are concentration pairs of the first or the second kind. The above lemma shows that for almost every x in A, the trace of R has full measure in $\{x\} \times \Sigma$. One can easily see that R is measurable for the product measure $\mathcal{H}^{N-1} \times \mathcal{H}^{N-1}$. Then, by Fubini's Theorem, the above conclusion means that R has full measure in $A \times \Sigma$ or, equivalently, that for almost every θ in Σ the trace of R has full measure in $A \times \{\theta\}$.

Lemma 12.25. *If A is an unrectifiable $(N-1)$-set, then almost every θ in Σ is a concentration direction of the first or the second kind for almost every x in A.*

A straightforward consequence of Lemmas 12.10, 12.17 and 12.23 is the following one, the last main result of this chapter.

Proposition 12.26. *If A is an unrectifiable $(N-1)$-set, then for almost every θ in Σ,*

$$\mathcal{H}^{N-1}(P_\theta(A)) = 0 .$$

Part III

EXISTENCE AND STRUCTURAL PROPERTIES
OF THE
MINIMAL SEGMENTATIONS
FOR THE
MUMFORD-SHAH MODEL

Chapter 13

PROPERTIES OF THE APPROXIMATING IMAGE IN THE MUMFORD-SHAH MODEL

In this chapter, we assume that a closed 1-set K with finite $\mathcal{H}^1$-measure has been associated with an image g as a possible "edge set". We shall not assume that K is minimal with respect to the Mumford-Shah functional because we wish to focus on the following question: Given K, what is to be said of $u = u_K$, where u is assumed to be the minimum point of the two-dimensional part of the Mumford-Shah energy,

$$I(u) = \int_{\Omega \setminus K} (|\nabla u|^2 + (u - g)^2)?$$

In Section 13.1, we explain some elementary properties satisfied by u, namely the elliptic equation $-\Delta u + u = g$, and we answer the following question: If K_n "tends to" K (in a sense which will be discussed), what can be said about the convergence of u_{K_n}, associated with K_n, towards $u = u_K$?

In Section 13.2, we look for the effect on the energy of u of a certain kind of modification of K: when some part of K has been erased. Then the new approximating function v satisfies $I(v) \geq I(u)$, and we give precise estimates on $I(v) - I(u)$ which relate this "jump of energy" to the geometry of K. Section 13.3 is devoted to an accurate estimate of the gradient of u, ∇u, as a function of the distance to K. When coupled with the "jump of energy" estimates of Section 13.2, this estimate will prove a basic tool to understand the geometry of minimal edge sets K.

1. Notation and preliminary lemmas.

Throughout this chapter, we shall work on open domains of the plane (denoted by B, Ω). Sometimes, these domains will be assumed to be "piecewise C^1", which means that their boundary is a finite union of C^1 Jordan curves. In particular, Ω can be thought of as a rectangle. We define the Mumford-Shah functional by

$$E(u, K) = \int_{\Omega \setminus K} |\nabla u|^2 + \int_{\Omega} (u - g)^2 + \mathcal{H}^1(K),$$

where the "image" g is a given bounded measurable function. In order to normalize the constants, we impose, without loss of generality, that $|g(x)| \leq \frac{1}{2}$ and that the diameter of Ω is less or equal than 1. The variable K denotes a closed subset of $\overline{\Omega}$ and u a function in the Sobolev Hilbert space $H^1(\Omega \setminus K)$ of the functions such that

$$\int_{\Omega \setminus K} (u(x)^2 + |\nabla u(x)|^2)dx < \infty.$$

It is important to understand the meaning of ∇u because the above notation may be ambiguous. Indeed, given K, we consider all functions u in $L^2(\Omega \setminus K)$ such that their distributional gradient is in $L^2(\Omega \setminus K)$.

We take the convention to extend u and ∇u on K by setting $u(x) = 0$ and $\nabla u = 0$ for any x in K. However, we keep the integration domain of ∇u to be $\Omega \setminus K$ in the above definition of $E(u, K)$. We do that in order to recall that ∇u is the distributional gradient of u on the complementary $\Omega \setminus K$ of K, and not on Ω.

We decompose the Mumford-Shah energy into the two-dimensional energy $I(u, K)$ and the one-dimensional Hausdorff measure, so that

$$E(u, K) = I(u, K) + \mathcal{H}^1(K).$$

The minimization approach for the Mumford-Shah functional can be organized in several ways, and one of them is certainly to fix K and thereafter to minimize with respect to u. If we do so, we are allowed to consider this minimum with respect to u as a K-depending new functional. We therefore write $E(K) = E(u_K, K)$ and $I(K) = I(u_K, K)$ where u_K is the function for which the value of $I(u, K)$ is minimal for a given K. When we fix K, we may consider E as a functional only depending on the function u. In this case (of a fixed K), we shall set without ambiguity

$$I(u) = \int_{\Omega \setminus K} (|\nabla u|^2 + (u - g)^2).$$

We shall also consider the restriction of this integral to some subdomain B. We then set (always in the case of a fixed K)

$$I_B(u) = \int_{B \setminus K} (|\nabla u|^2 + (u - g)^2).$$

If B is an open set with piecewise C^1 boundary, we denote by $\nu(x)$ the exterior normal to B at a point x of ∂B. This normal is well defined at all points but a finite set and if u is a C^1 function on $\overline{B}$, we set, as usual,

$$\frac{\partial u}{\partial n} = \nabla u . \nu,$$

which is the derivative of u in the direction normal to ∂B. Throughout the next chapters, we shall make an extensive use of the following classical existence and regularity results for elliptic equations [GilT]. The next lemma solves the problem of finding u minimizing $I(u)$ when K is fixed, since we can take as a particular case $B = \Omega \setminus K$.

Neumann Lemma 13.1. *Let $|g| \leq \frac{1}{2}$ be a measurable bounded function, defined on an open domain B. Consider the function u defined as the unique minimizer of the convex functional on $H^1(B)$,*

$$I(u) = \int_B (|\nabla u|^2 + (u - g)^2).$$

Then u is a C^1 function on B satisfying

(13.2) $$-\Delta u + u = g \qquad on \quad B$$

in the distributional sense. If, in addition, ∂B is piecewise C^1, then u is C^1 up to the boundary of B, excepting the finite set of singular points of ∂B, and satisfies the Neumann boundary condition

(13.3) $$\frac{\partial u}{\partial n} = 0 \quad on \quad \partial B.$$

If u is as in the preceding lemma, it has a useful boundedness property, known as "variational maximum principle", which is a consequence of the Stampacchia Lemma.

Stampacchia Lemma 13.4. *Let B be an open domain, u a function in $H^1(B)$ and a a real number. Then $u^a = max(a, u) \in H^1(B)$ and*

$$\int_B |\nabla u^a|^2 \leq \int_B |\nabla u|^2.$$

The proof of the Stampacchia Lemma, which is intuitively obvious as we shall now see, can be found, for example, in Brezis, [Bre], Kinderlehrer-Stampacchia [KiSt] and Gilbarg-Trüdinger [GilT]. In the case where u is C^1, one can base the proof on the straightforward remark that $\nabla u^a(x) = 0$ on $\{x, \quad u(x) > a\}$ and $\nabla u^a(x) = \nabla u(x)$ elsewhere.

Variational Maximum Principle 13.5. *Let u be as in Neumann Lemma 13.1. Then $|u(x)| \leq \frac{1}{2}$.*

Proof. Set $\tilde{u}(x) = Max(-\frac{1}{2}, Min(\frac{1}{2}, u(x)))$. Then $|\tilde{u} - g| \leq |u - g|$ and, by Stampacchia Lemma, $\tilde{u}$ belongs to $H^1(\Omega)$ and satisfies $I(\tilde{u}) \leq I(u)$. Since the functional $I(u)$ is strictly convex, its minimum is unique and therefore $u = \tilde{u}$ almost everywhere. $\square$

We now consider a fixed domain Ω and study the continuity properties of the functional $I(K)$, depending of the set variable K, with respect to the Hausdorff distance.

Semicontinuity Lemma 13.6. *Assume that $K_n \to K$ for the Hausdorff distance, then $I(K) \leq \liminf I(K_n)$. In addition, if $I(K) = \lim_n I(K_n)$, then*

$$(13.7) \qquad u_{K_n} \to u_K, \quad \nabla u_{K_n} \to \nabla u_K \quad \text{in } L^2(\Omega \setminus K).$$

Proof. Set for simplicity $u_n = u_{K_n}$. The sequence of the functions u_n, considered as defined on all of Ω in the sense explained above (extended by zero on K_n), is certainly bounded in $L^2(\Omega)$. Indeed

$$\int_\Omega (u - g)^2 \leq I(u_n) \leq I(0) = \int_\Omega g^2 < \infty.$$

So, after replacing u_n with a suitable subsequence, we can assume that it converges to a weak limit u in $L^2(\Omega)$. In the same way, we can assume that ∇u_n weakly converges to a vector function v in $L^2(\Omega)$. We claim that $v = \nabla u$ on the open set $\Omega \setminus K$. In fact, take any test function φ with support in $\Omega \setminus K$. For n large we have by compactness

$$\text{support}(\varphi) \subset \Omega \setminus K_n$$

and, consequently,

$$\int u_n \nabla \varphi = - \int (\nabla u_n) \varphi.$$

By definition of the weak convergence, the left-hand side of the above relation converges to $\int u \nabla \varphi$ and the right one to $- \int v \varphi$. By the arbitrariness of φ, we obtain $\nabla u = v$ on $\Omega \setminus K$, as claimed. The first part of the thesis follows because, from the lower semicontinuity of the L^2-norm for the weak topology in L^2, we have

$$I(K) \leq I(u, K) \leq \liminf I(u_n, K_n).$$

When $I(K) = \lim_n I(K_n)$, the above inequalities become equalities, which implies that $u = u_K$ and that the L^2-norms of u_n and ∇u_n converge to the L^2-norms of u and ∇u. This makes (13.7) hold for the whole sequence, since the weak limit point is actually a strong one and does not depend on the particular subsequence. □

Corollary 13.8: Strong Convergence of Approximates. *Two cases for which (13.7) obviously holds are when*

$$\forall n, \quad K \subset K_n \quad \text{or when}$$

$$E(K_n) \to \text{Inf}_K E(K) \quad \text{and} \quad \mathcal{H}^1(K) \leq \liminf \mathcal{H}^1(K_n).$$

The first relation holds because the functional $I(K)$ is nonincreasing with respect to K. In other terms, if $K \subset K'$, then $I(K) \geq I(K')$.

2. Green Formulas.

*In this section, as a direct application of the Neumann Lemma 13.1
and the Green Formula 13.9 below, we shall show how to calculate the
jump of the energy $I(u) - I(v)$ when we replace a function u defined on
$\Omega \setminus K$ by another.*

Let us recall the well known Green Formula, whose proof essentially
is an integration by parts. A proof can be found (e.g.) in [Bre, KiSt].

Green Formula 13.9. *Let B be an open set of the plane such that ∂B
is piecewise C^1. Let u and v be two functions in $C^1(\overline{B})$ and assume that
Δv is a bounded function on B. Then*

$$\int_{\partial B} u \frac{\partial v}{\partial n} = \int_B (\nabla u \nabla v + u \Delta v).$$

Green Lemma 13.10. *Let B be an open domain with piecewise C^1
boundary and u and v be two functions of $C^1(\Omega \setminus B)$ satisfying (13.2) on
$\Omega \setminus B$. Then*

$$I_{\Omega \setminus B}(u) - I_{\Omega \setminus B}(v) = \int_{\partial(\Omega \setminus B)} (u - v) \frac{\partial(u + v)}{\partial n}.$$

Proof. By Green Formula 13.9,

$$I_{\Omega \setminus B}(u) - I_{\Omega \setminus B}(v) = \int_{\Omega \setminus B} (|\nabla u|^2 - |\nabla v|^2 + (u - g)^2 - (v - g)^2)$$

$$= \int_{\Omega \setminus B} (\nabla(u - v)\nabla(u + v) + (u - v)(u + v - 2g))$$

$$= \int_{\Omega \setminus B} (u - v)(-\Delta(u + v) + u + v - 2g)$$

$$+ \int_{\partial(\Omega \setminus B)} (u - v) \frac{\partial(u + v)}{\partial n}.$$

This achieves the proof because u and v satisfy (13.2). $\square$

In order to be applied, the Green Formula and the preceding Green
Lemma 13.10 ask that the boundaries of the considered domain are piecewise C^1. Weak versions, with nonsmooth boundaries can be established,
but they need some technical preliminaries. Therefore, since we wish to
be able to apply the Green Formula on $B \setminus K$, which is not smooth, because K is not, we shall establish an approximation method to replace
K by a smoother set, K_ε.

Lemma 13.11: Approximation of K. *Let K be a compact 1-set of Ω. Then there exists a sequence $(K_\varepsilon) \supset K$ of sets such that the boundaries ∂K_ε are piecewise C^1, $K_\varepsilon \to K$ for the Hausdorff distance and $\mathcal{L}_2(K_\varepsilon) \to 0$ as $\varepsilon \to 0$.*

Proof. Take a finite covering of K made of disks, $\mathcal{B}_\varepsilon$, such that $\rho(\mathcal{B}_\varepsilon) \le \varepsilon$ and $|\mathcal{B}_\varepsilon|_1 \le 2\mathcal{H}^1(K)$. This is possible because of the definition (§6.1) and the compactness of K. Set $K_\varepsilon = \bigcup_{D \in \mathcal{B}_\varepsilon} D$. Then $\mathbf{d}(K_\varepsilon, K) \to 0$ and

$$\mathcal{L}_2(K_\varepsilon) \le \frac{\pi}{4} \sum_{D \in \mathcal{B}_\varepsilon} (\mathrm{diam}(D))^2 \le \frac{\pi \varepsilon}{2} |\mathcal{B}_\varepsilon|_1.$$

$\square$

If we approximate K by K_ε, we also get an approximation of $u = u_K$ by setting $u_\varepsilon = u_{K_\varepsilon}$. Indeed, by the above Semicontinuity Lemma (13.6), $I(u_\varepsilon)$ tends to $I(u)$, u_ε tends to u in $L^2(\Omega)$ and ∇u_ε tends to ∇u in $L^2(\Omega \setminus K)$. (Recall that by an above-mentioned convention, we extend ∇u_ε on $\Omega \setminus K$ by setting $\nabla u_\varepsilon = 0$ on K_ε.) In order to extend the Green Formula, we are interested in having as much information as possible about the convergence of the restriction of u_ε and ∇u_ε to the boundary of disks.

Definition: reliable sets. *We shall say that an open set B of Ω is reliable with respect to $u = u_K$ if there is a sequence K_ε approximating K as defined in Lemma 13.11 such that $u_\varepsilon = u_{K_\varepsilon}$ and ∇u_ε respectively converge to u and ∇u in $L^2(\partial B)$. (By our convention, u_ε, u, ∇u and ∇u_ε are pointwise well defined on Ω.) We shall say that B is reliable with respect to K if $\mathcal{H}^1(K_\varepsilon \cap \partial B) \to 0$ when $\varepsilon \to 0$.*

Proposition 13.12. *Let K be a compact 1-set of Ω and $x \in \Omega$. Then for almost every r such that $D(x, r) \subset \Omega$, $D(x, r)$ is reliable with respect to $u = u_K$ and K.*

Proof. Fix a sequence K_ε as specified in Lemma 13.11. By the semicontinuity Lemma 13.6, u_ε and ∇u_ε converge in $L^2(\Omega \setminus K)$ (and therefore in $L^2(\Omega)$ by our convention). Using Fubini's Theorem, we deduce that for a fixed x and almost every r, the disk $D(x, r)$ is reliable with respect to u_K. Since $\mathcal{L}_2(K_\varepsilon) \to 0$, the same argument shows that $D(x, r)$ is reliable with respect to K for almost every r. $\square$

Green Formula with Nonsmooth Boundary 13.13. *Let K be a closed 1-set of Ω, $u = u_K$, B an open subset of Ω with piecewise C^1 boundary, reliable with respect to (u, K). Then*

$$\int_{\partial B \setminus K} u \frac{\partial u}{\partial n} = \int_{B \setminus K} (|\nabla u|^2 + u \Delta u),$$

$$\int_{\partial B \setminus K} \frac{\partial u}{\partial n} = \int_{B \setminus K} \Delta u.$$

Proof. Let K_ε, u_ε be approximates of K and u as defined in (13.11). Then, taking into account that u_ε is extended by zero on K_ε, using $\frac{\partial u_\varepsilon}{\partial n} = 0$ on ∂K_ε (by Neumann Lemma 13.1) and applying the Green Formula (13.9) on $B \setminus K_\varepsilon$, we obtain

$$(13.14) \quad \int_{\partial B \setminus K} u_\varepsilon \frac{\partial u_\varepsilon}{\partial n} = \int_{\partial B \setminus K_\varepsilon} u_\varepsilon \frac{\partial u_\varepsilon}{\partial n} = \int_{B \setminus K_\varepsilon} (|\nabla u_\varepsilon|^2 + u_\varepsilon \Delta u_\varepsilon).$$

Let us pass to the limit in the preceding relation. Since B is reliable, we have that

$$\int_{\partial B \setminus K} u_\varepsilon \frac{\partial u_\varepsilon}{\partial n} \to \int_{\partial B \setminus K} u \frac{\partial u}{\partial n}.$$

In addition, by the Corollary 13.8 and the Semicontinuity Lemma 13.6, $u_\varepsilon \to u$ and $\nabla u_\varepsilon \to \nabla u$ in $L^2(B \setminus K)$, namely in $L^2(B)$ since the functions are extended on K by 0. Taking the convention to extend Δu_ε by $-g$ on K_ε and Δu by $-g$ on K, we deduce by (13.2), $\Delta u_\varepsilon = u_\varepsilon - g \to u - g = \Delta u$ in $L^2(B)$. We conclude that the last integral of (13.14) converges to

$$\int_{B \setminus K} (|\nabla u|^2 + u \Delta u),$$

which achieves the proof. $\square$

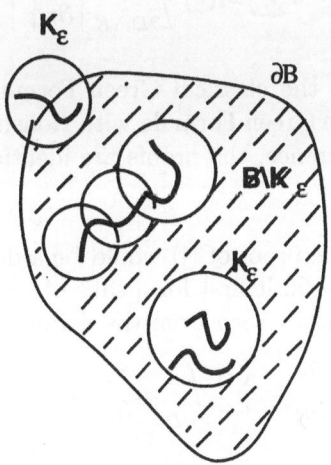

Figure 13.1 : Green Formula with nonsmooth boundary

We now consider the problem of estimating Green integrals like

$$\int_{\partial B} u \frac{\partial v}{\partial n},$$

appearing in the last statements. In the following, B will appear as a finite union $B = \cup_i D_i$ of small open sets (mainly disks and squares) and we shall have easy estimates of the Green integrals on each ∂D_i. So we need a technical tool which allows to pass from separate estimates of the Green integrals to the global one. This will be fulfilled in the next lemma, which will therefore prove extremely useful in the remainder of this book.

Lemma 13.15: Localization of Green Formula.

(i) Let $\{D_i\}_{i=1,2,...,m}$ be bounded open subsets of $\mathbb{R}^2$ with connected piecewise C^1 boundary and $B = \cup_i D_i$. Assume that B also has a piecewise C^1 boundary and that no D_i is contained in any D_j with $i \neq j$. Let u be a bounded function defined on ∂B and satisfying $|u| \leq \frac{1}{2}$ and $v \in H^1(B)$ satisfying $|\Delta v| \leq 1$. Then

$$(13.16) \qquad \left| \int_{\partial B} u \frac{\partial v}{\partial n} \right| \leq 2 \sum_i \omega_i(u) \int_{\partial D_i} \left| \frac{\partial v}{\partial n} \right| + \frac{1}{2} \mathcal{L}_2(B),$$

where $\omega_i(u) = \sup_{\partial D_i} u - \inf_{\partial D_i} u$ denotes the oscillation of u on ∂D_i.

(ii) Let K be a closed set of $\mathbb{R}^2$, let the D_i and B be as above and assume that they are reliable with respect to $u = u_K$. Assume that v is a function defined on B and satisfying $|v| \leq \frac{1}{2}$. Then

$$(13.17) \qquad \left| \int_{\partial B \backslash K} v \frac{\partial u}{\partial n} \right| \leq 2 \sum_i \omega_i(v) \int_{\partial D_i \backslash K} \left| \frac{\partial u}{\partial n} \right| + \frac{1}{2} \mathcal{L}_2(B).$$

Proof. The proof of (i) uses the classical Green Formula applied to v, while the proof of (ii) uses the Green Formula with Nonsmooth Boundary (13.13). Except for this difference, the proofs are identical and we shall prove (i).

If the D_i were a partition, the proof of (i) would be a direct application of the Green Formula. Let us do it first for a sake of graduate difficulty. In this particular case, it is easily seen from the definition of $\frac{\partial v}{\partial n}$ that

$$\int_{\partial B} u \frac{\partial v}{\partial n} = \sum_i \int_{\partial D_i} u \frac{\partial v}{\partial n}.$$

Choose values $u_i \in u(\partial D_i)$. Then we can write

$$\int_{\partial B} u \frac{\partial v}{\partial n} = \sum_i \int_{\partial D_i} (u - u_i) \frac{\partial v}{\partial n} + \sum_i u_i \int_{\partial D_i} \frac{\partial v}{\partial n}.$$

The announced relation (i) easily follows by applying the Green Formula $\int_{\partial D_i} \frac{\partial v}{\partial n} = \int_{D_i} \Delta v$ and the assumed bounds on u and Δv.

The general case is a little bit tricky but in the very same line. We shall work by induction, and we reorder the indices i in such a way that $(j \le i) \Rightarrow (\omega_j(u) \le \omega_i(u))$. Let $B^1 = \bigcup_{i>1} D_i$ and set

$$I_i = \{j \,|\, \partial D_j \cap \partial D_i \neq \emptyset \quad \text{and} \quad j \le i\}.$$

Call again, for each i, u_i a fixed value assumed by u on ∂D_i. Denote by $\tilde{U}$ the set of functions $\tilde{u}$ defined on $\bigcup_i \partial D_i$ such that

$$(13.18) \qquad \tilde{u}(\partial D_i) \subset u(\partial D_i) \cup \left(\bigcup_{j \in I_i} \{u_j\} \right).$$

We have

$$(13.19) \qquad \int_{\partial B} u \frac{\partial v}{\partial n} = \int_{\partial B \cap \overline{D}_1} u \frac{\partial v}{\partial n} + \int_{\partial B \setminus \overline{D}_1} u \frac{\partial v}{\partial n}.$$

Let us introduce the oscillation of u on ∂D_1 by writing

$$(13.20) \qquad \int_{\partial B \cap \overline{D}_1} u \frac{\partial v}{\partial n} = \int_{\partial B \cap \overline{D}_1} (u - u_1) \frac{\partial v}{\partial n} + \int_{\partial B \cap \overline{D}_1} u_1 \frac{\partial v}{\partial n}.$$

In order to estimate the second term of this relation by a Green Formula, we use the topological relation

$$(13.21) \qquad \partial(\overline{D}_1 \setminus B^1) = (\partial B \cap \overline{D}_1) \cup (\partial B^1 \cap \overline{D}_1).$$

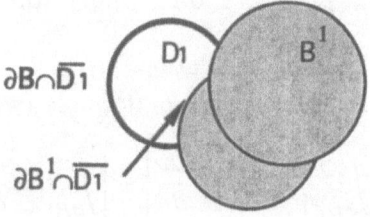

Figure 13.2 : Localization of Green Formula

This relation is true because if A and B are two closed sets, then $\partial(A \cap B) = (\partial A \cap B) \cup (A \cap \partial B)$. Applying this relation we obtain

$$\partial(\overline{D}_1 \setminus B^1) = \partial(\overline{D}_1 \cap (B^1)^c) = (\partial \overline{D}_1 \cap (B^1)^c) \cup (\overline{D}_1 \cap \partial B^1),$$

and we end the proof of the formula (13.21) by noting that

$$\partial \overline{D}_1 \cap (B^1)^c = \partial \overline{D}_1 \setminus B^1 = \overline{D}_1 \setminus (D_1 \cup B_1) = (\overline{D}_1 \cap \overline{B}) \setminus B = \overline{D}_1 \cap \partial B.$$

Using Green Formula, taking into account the orientation of the boundaries and the fact that u_1 is a constant, we obtain from (13.21)

$$(13.22) \qquad \int_{\overline{D}_1 \setminus B^1} u_1 \Delta v \, dx = \int_{\partial B \cap \overline{D}_1} u_1 \frac{\partial v}{\partial n} - \int_{\partial B^1 \cap \overline{D}_1} u_1 \frac{\partial v}{\partial n}.$$

By combining the relations (13.19), (13.20), (13.22) we get

$$(13.23) \qquad \begin{aligned} \int_{\partial B} u \frac{\partial v}{\partial n} &= \int_{\partial B \cap \overline{D}_1} (u - u_1) \frac{\partial v}{\partial n} + \int_{\partial B^1 \cap \overline{D}_1} u_1 \frac{\partial v}{\partial n} + \\ &\quad + \int_{\overline{D}_1 \setminus B^1} u_1 \Delta v \, dx + \int_{\partial B \setminus \overline{D}_1} u \frac{\partial v}{\partial n}. \end{aligned}$$

So, if we define $\tilde{u}_2$ by

$$\tilde{u}_2 = u_1 \qquad \text{on} \quad \partial B^1 \cap \overline{D}_1$$

$$\tilde{u}_2 = u \qquad \text{on} \quad \left(\bigcup_i \partial D_i \right) \setminus \overline{D}_1,$$

it is easily seen that if u is in $\tilde{U}$, then $\tilde{u}_2 \in \tilde{U}$. Indeed, if $x \in \partial B^1 \cap \overline{D}_1$, then for some $i > 1$, $x \in \partial D_i$ and since ∂D_i meets $\overline{D}_1$ and, by assumption, $D_i \not\subseteq D_1$, then ∂D_i meets ∂D_1. Thus $i \in I_1$. Using this notation, we can rewrite (13.23) as

$$\int_{\partial B} u \frac{\partial v}{\partial n} = \int_{\partial B \cap \overline{D}_1} (u - u_1) \frac{\partial v}{\partial n} + \int_{\partial B^1} \tilde{u}_2 \frac{\partial v}{\partial n} + \int_{\overline{D}_1 \setminus B^1} u_1 \Delta v \, dx.$$

We now set $\tilde{u}_1 = u$ and use the relations $|u_1| \leq \frac{1}{2}$, $|\Delta v| \leq 1$ and $\partial B \cap \overline{D}_1 \subset \partial D_1$ to obtain the following recursion formula (where $\tilde{u}_1 = u \in \tilde{U}$)

$$\left| \int_{\partial B} \tilde{u}_1 \frac{\partial v}{\partial n} \right| \leq \int_{\partial D_1} \left| (\tilde{u}_1 - u_1) \frac{\partial v}{\partial n} \right| + \left| \int_{\partial B^1} \tilde{u}_2 \frac{\partial v}{\partial n} \right| + \frac{1}{2} \mathcal{L}(\overline{D}_1 \setminus B^1).$$

Therefore, by iterating the above argument and defining recursively $\tilde{u}_i \in \tilde{U}$,

$$\left| \int_{\partial B} u \frac{\partial v}{\partial n} \right| \leq \sum_i \int_{\partial D_i} \left| (\tilde{u}_i - u_i) \frac{\partial v}{\partial n} \right| + \frac{1}{2} \mathcal{L}(B) .$$

Let us now estimate the upper bound of $|\tilde{u}_i - u_i|$ on ∂D_i. If $\tilde{u}_i(x) \in u(\partial D_i)$ then $|\tilde{u}_i(x) - u_i| \leq \omega_i(u)$. If $\tilde{u}_i(x) \in \bigcup_{j \in I_i} \{u_j\}$ then, taking $y \in \partial D_i \cap \partial D_j$, one has

$$\begin{aligned} |\tilde{u}_i(x) - u_i| = |u_j - u_i| &\leq |u_j - u(y)| + |u(y) - u_i| \\ &\leq \omega_j(u) + \omega_i(u) \\ &\leq 2\omega_i(u) . \end{aligned}$$

$\square$

3. Interior regularity estimates.

We shall now prove a very useful estimate on the gradient of a function which satisfies an elliptic equation like (13.2) on a disk, in terms of the distance from the boundary. To this aim we begin by pointing out an integral identity enjoyed by the functions of two variables which are harmonic on a disk. For every (x, y) of the plane, we consider the unit vectors $n = \frac{(x,y)}{(x^2+y^2)^{\frac{1}{2}}}$ and $\theta = \frac{(-y,x)}{(x^2+y^2)^{\frac{1}{2}}}$ and we set for any smooth function $v(x, y)$,

$$\frac{\partial v}{\partial \rho} = \nabla v.n, \qquad \frac{\partial v}{\partial \tau} = \nabla v.\theta.$$

We also set

$$\frac{\partial^2 v}{\partial \rho^2} = D^2 v(n, n), \qquad \frac{\partial^2 v}{\partial \tau^2} = D^2 v(\theta, \theta)$$

and

$$\frac{\partial^2 v}{\partial \tau \partial \rho} = D^2 v(\theta, n).$$

With these definitions, it is easily checked that

$$\Delta v = \frac{\partial^2 v}{\partial \rho^2} + \frac{\partial^2 v}{\partial \tau^2},$$

$$\frac{\partial}{\partial \rho} \left(\frac{\partial v}{\partial \rho} \right) = \frac{\partial^2 v}{\partial \rho^2} \qquad \frac{\partial}{\partial \rho} \left(\frac{\partial v}{\partial \tau} \right) = \frac{\partial^2 v}{\partial \rho \partial \tau},$$

and

$$\frac{\partial}{\partial \tau} \left(\frac{\partial v}{\partial \rho} \right) = \frac{\partial^2 v}{\partial \rho \partial \tau} + \rho^{-1} \frac{\partial v}{\partial \tau}, \qquad \frac{\partial}{\partial \tau} \left(\frac{\partial v}{\partial \tau} \right) = \frac{\partial^2 v}{\partial \tau^2} - \rho^{-1} \frac{\partial v}{\partial \rho}.$$

Lemma 13.24. *Let v be a harmonic function defined on a closed disk D, that is, a C^1 function on D satisfying $\Delta v = 0$. Then*

$$(13.25) \qquad \int_{\partial D} \left(\frac{\partial v}{\partial \rho} \right)^2 d\mathcal{H}^1 = \int_{\partial D} \left(\frac{\partial v}{\partial \tau} \right)^2 d\mathcal{H}^1.$$

Proof.

Let R be the radius of D. For ρ in the interval $(0,R)$ define

$$i_1(\rho) = \rho^{-1} \int_{\partial D_\rho} \left(\frac{\partial v}{\partial \rho} \right)^2 d\mathcal{H}^1$$

$$i_2(\rho) = \rho^{-1} \int_{\partial D_\rho} \left(\frac{\partial v}{\partial \tau} \right)^2 d\mathcal{H}^1.$$

An easy computation shows that

$$\frac{d}{d\rho} i_1(\rho) = 2\rho^{-1} \int_{\partial D_\rho} \frac{\partial^2 v}{\partial \rho^2} \frac{\partial v}{\partial \rho} d\mathcal{H}^1$$

and an analogous one, combined with an integration by parts, gives

$$\frac{d}{d\rho} i_1(\rho) = 2\rho^{-1} \int_{\partial D_\rho} \frac{\partial^2 v}{\partial \rho \partial \tau} \frac{\partial v}{\partial \tau} d\mathcal{H}^1 =$$

$$= 2\rho^{-1} \int_{\partial D_\rho} \left(\frac{\partial}{\partial \tau} \left(\frac{\partial v}{\partial \rho} \right) - \rho^{-1} \frac{\partial v}{\partial \tau} \right) \frac{\partial v}{\partial \tau} =$$

$$= -2\rho^{-1} \int_{\partial D_\rho} \frac{\partial v}{\partial \rho} \frac{\partial}{\partial \tau} \left(\frac{\partial v}{\partial \tau} \right) - 2\rho^{-2} \int_{\partial D_\rho} \left(\frac{\partial v}{\partial \tau} \right)^2 =$$

$$= -2\rho^{-1} \int_{\partial D_\rho} \frac{\partial v}{\partial \rho} \left(\frac{\partial^2 v}{\partial \tau^2} - \rho^{-1} \frac{\partial v}{\partial \rho} \right) - 2\rho^{-2} \int_{\partial D_\rho} \left(\frac{\partial v}{\partial \tau} \right)^2 =$$

$$= -2\rho^{-1} \int_{\partial D_\rho} \frac{\partial^2 v}{\partial \tau^2} \frac{\partial v}{\partial \rho} d\mathcal{H}^1 + 2\rho^{-1}(i_1(\rho) - i_2(\rho)).$$

Therefore we have

$$(13.26) \quad \frac{d}{d\rho}(i_1(\rho) - i_2(\rho)) = 2\rho^{-1} \int_{\partial D_\rho} \Delta v \frac{\partial v}{\partial \rho} d\mathcal{H}^1 - 2\rho^{-1}(i_1(\rho) - i_2(\rho)).$$

The last formula implies that

$$\frac{d}{d\rho}(\rho^2(i_1(\rho) - i_2(\rho))) = 0$$

and consequently the equality

$$i_1(\rho) = i_2(\rho)$$

must hold for every $\rho \leq R$. For $\rho = R$ we obtain (13.25).

□

The above lemma permits us to prove an interior estimate for the gradient of solutions of elliptic equations like (13.2) on a disk.

Lemma 13.27. *Let D be a closed disk with radius $R \leq 1$ and v be a C^1 function defined on D satisfying $|\Delta v| \leq 1$. Assume that*

$$C^2 = \frac{4}{\pi} \int_{\partial D} \left(\frac{\partial v}{\partial \tau}\right)^2 d\mathcal{H}^1 < \infty.$$

Then for every x in D there exists a constant C' such that

$$|\nabla v(x)| \leq \frac{C'}{\sqrt{d(x, \partial D)}}.$$

Proof. We begin by proving the lemma in the case where v is harmonic. Note that Lemma 13.24 implies that

(13.28) $$C^2 = \frac{2}{\pi} \int_{\partial D} |\nabla v|^2 d\mathcal{H}^1.$$

Let R be the radius of D. For ρ in the interval $(0,R)$ consider the function

$$\rho \to \rho^{-1} \int_{\partial D_\rho} |\nabla v|^2 d\mathcal{H}^1.$$

Since v is assumed to be harmonic, it is easily seen the function $|\nabla v|^2$ is subharmonic (i.e., $\Delta(|\nabla v|^2) \geq 0$), so that, as a consequence of Green Lemma, the above function of ρ is nondecreasing. This implies that

$$\forall \rho \in (0, R) : \int_{\partial D_\rho} |\nabla v|^2 d\mathcal{H}^1 \leq \int_{\partial D} |\nabla v|^2 d\mathcal{H}^1.$$

Thus, if we set $d = d(x, \partial D)$ and we denote by A the annulus with the same center as D and radii $R - 2d$ and R, we have

$$\int_A |\nabla v|^2 dx \le 2d \int_{\partial D} |\nabla v|^2 d\mathcal{H}^1.$$

Let B be the ball centered at x with radius d. Clearly $B \subset A$. By the subharmonicity of $|\nabla v|^2$ and (13.28) we finally get

$$|\nabla v(x)|^2 \le \mathcal{L}(B)^{-1} \int_B |\nabla v|^2 dx \le (\pi d^2)^{-1} \int_A |\nabla v|^2 dx$$
$$\le \frac{2}{\pi} d^{-1} \int_{\partial D} |\nabla v|^2 d\mathcal{H}^1 \le C^2 d^{-1}.$$

Let us now pass to the proof in the case where u is not harmonic but satisfies $|\Delta u| \le 1$ on a disk with radius less than 1. We just have to subtract from u a function w with the same Laplacian and equal to zero on the boundary of the disk. By a classical result about the regularity of solutions of elliptic equations [Bre, KiSt, GilT], such a function has a gradient bounded by an absolute constant C''. So we can apply the just proved estimate on the gradient of a harmonic function to the difference $v = u - w$, which is harmonic. Then, since $d(x, \partial D) \le 1$, u satisfies the conclusion of Lemma (13.27), with a constant $C' = C + C''$. $\square$

Chapter 14

SMALL OSCILLATION COVERINGS AND THE EXCISION METHOD

*In this chapter, we develop the main general tool used for proving reg-
ularity properties of the minimizers of the Mumford-Shah functional. It
will be called Excision Method and consists in showing that, whenever
the set K is too "ragged" inside some disk D, and whenever u is smooth
enough around each piece of K, then the excision operation, consisting
of removing from the edge set K the points contained in D, decreases the
value of the functional E(K).*

*This kind of information will have a straightforward use: since the
functional E(K) cannot decrease when the edge set K is a minimum,
we shall conclude that K is not too "ragged", whatever the disk D is,
and regularity properties for K will follow. Let us now see roughly which
kind of raggedness condition we shall consider. We shall assume that
the edge set can be covered by a family of sets D_i on whose boundary,
∂D_i, the function $u = u_K$ does not oscillate much. Such coverings will be
called* small oscillation coverings. *We shall prove that when one can cover
the edge set K with a small oscillation covering made of sets D_i whose
size is small enough with respect to the diameter of D, then the excision
operation works successfully and decreases the Mumford-Shah energy.*

*Several applications of this general excision principle will be done in
the next two chapters. The "concrete properties" of the minimizers which
will be proved therein will in fact be straightforward applications of the
excision method.*

1. The excision method.

In this chapter, we shall consider the following generic situation. K is
a closed 1-set, $u = u_K$ is associated with K as in the preceding chapter.
$D(s)$ is a disk with a fixed center belonging to K whose radius s varies
beween $\frac{R}{2}$ and $R \le 1$ is a given positive constant.

Definition 14.1. *We say that $K(s)$ is obtained by excision of $D(s)$ from
K if*

$$(14.2) \qquad K(s) = (K \setminus D(s)) \cup T(s),$$

*where $T(s)$ is a closed subset of $\partial D(s)$ containing $K \cap \partial D(s)$ which can be
empty, equal to the whole $\partial D(s)$, or a finite union of closed arcs, each one*

of which meets K and such that, if two points are in the same connected component of K and in $\partial D(s)$ they belong to a same arc. We choose a solution v_s of

(14.3)
$$-\Delta v_s + v_s = g \quad in \quad D(s)$$
$$v_s = u \quad on \quad \partial D(s) \setminus T(s).$$

Note that v_s is determined by its values on $T(s)$, which we are free to choose in a convenient way. We set

$$u_s = u \quad in \quad \Omega \setminus D(s)$$
$$u_s = v_s \quad on \quad D(s).$$

If u_s is as above, we say that $(K(s), u(s))$ is obtained by excision of $D(s)$ from (K, u).

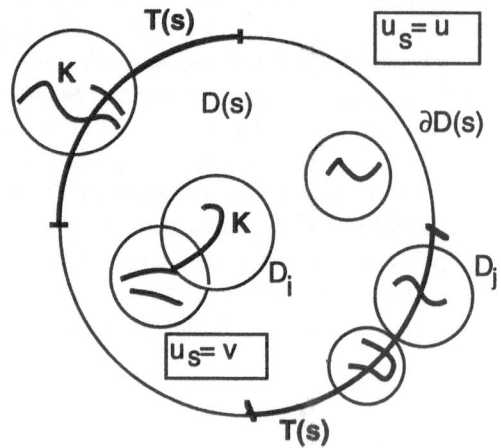

Figure 14.1 : Excision method

The restrictions on the choice of $T(s)$ have been made in order to ensure that the number of the connected components of $K(s)$ is smaller or equal to that of K. Indeed, if $\partial D(s) \cap K = \emptyset$, then since $D(s)$ contains a point of K, at least one connected component of K must be contained in $D(s)$ and is therefore removed by the excision operation. So, even if we add the circle $\partial D(s)$ to K, the number of connected components cannot increase. In the other case we are only allowed to add arcs which meet K, so that the number of components does not increase from K to $K \cup T(s)$.

The whole geometric situation of excision is summarized in Figure 14.1. After excision, all pieces of K inside $D(s)$ disappear but some new "cut", $T(s)$, has been added to the segmentation. Now, u has been replaced by a smooth function v_s inside $D(s)$. Let us consider a quasiminimality property for segmentations which is naturally associated with excision.

Definition 14.4 (Property (M)). *We shall say that a set K satisfies the minimality assumption (M) if $E(K(s)) \geq E(K)$ for every set $K(s)$ obtained from K by excision, according to Definition 14.1.*

It is important to notice that, since $R \leq 1$, (M) implies

$$(14.5) \qquad \int_{D(R)} |\nabla u|^2 \leq CR.$$

In fact, if (14.5) does not hold with $C = 3\pi$, we can set $K(R) = (K \setminus D(R)) \cup \partial D(R)$ and, denoting by w the function equal to u on $\Omega \setminus D(R)$ and to zero on $D(R)$, we have, by (M), $E(K(R), w) \geq E(K, u)$, which implies $I(w) \geq I(u) - 2\pi R$. Since it is easily seen that

$$I(w) = I(u) - \int_{D(R)} |\nabla u|^2 + \int_D (g^2 - (u-g)^2)$$
$$< I(u) - CR + \pi R^2 = I(u) - 2\pi R,$$

we get in an obvious contradiction. By the same argument, we see that

$$(14.6) \qquad \mathcal{H}^1(K \cap D(R)) \leq 3\pi R.$$

2. Small Oscillation Coverings.

As we shall see, the minimality property (M) implies the nonexistence of any *small oscillation covering* of $K \cap D(R)$. In the remainder of this chapter, we assume that K satisfies the minimality assumption (M). We denote by $(D_i)_{i \in I}$ a piecewise C^1 covering of $K \cap D(R)$. (We say that a covering is piecewise C^1 if each ∂D_i is a finite union of piecewise C^1 Jordan curves.) We set $p_i = \mathcal{H}^1(\partial D_i)$ and $B = \bigcup_i D_i$. In the remainder of this chapter, whenever we write "C" in an inequality, this means that C is a constant, depending on no parameter and which could be explictely computed (like the C appearing in (14.5), which is 3π). We avoid explicit constants in order to make the proofs as clear as possible. In the same way, we shall write "$C(\nu)$" for a constant only depending on ν, and so on.

Definition 14.7. *We shall say that a piecewise C^1 covering, consisting of connected and simply connected sets $(D_i)_i$, is a small oscillation covering of $K \cap D(R)$, with constants μ and ν, if for each i, D_i is reliable with respect to u and K, ∂D_i meets K at at most one point,*

$$(14.8) \qquad \int_{\partial D_i} |\nabla u| \le \nu p_i^{\frac{1}{2}}$$

and

$$(14.9) \qquad \sum_i p_i \le \mu R.$$

We shall sometimes find it convenient to replace (14.9) by some stronger variants as, for instance,

$$(14.10) \qquad \sum_i p_i \le \mu \mathcal{H}^1(K \cap D(\frac{R}{2})).$$

Since K satisfies the minimality assumption (M), (14.10) actually implies (14.9), with the value of μ replaced by $\frac{3}{2}\pi\mu$. Indeed, (M) implies (14.6) which, combined with (14.10), gives

$$\sum_i p_i \le \frac{3}{2}\pi\mu R.$$

In order to simplify the notation, we shall denote the oscillation of a function w on the set ∂D_j by

$$\omega_j(w) = \sup_{\partial D_j} w - \inf_{\partial D_j} u.$$

Since ∂D_i meets K at at most one point, by our topological assumptions on the sets D_i, we have that $\partial D_i \setminus K$ is connected and therefore

$$(14.11) \qquad \forall i \in I, \quad \omega_i(u) \le \int_{\partial D_i} |\nabla u| \le \nu p_i^{\frac{1}{2}},$$

which justifies the name "small oscillation covering". The main "excision" lemmas of this chapter yield precise estimates of the increase of Mumford-Shah energy by excision, that is,

$$E(u_s, K(s)) - E(u, K) = I_D(v_s) - I_D(u) + \mathcal{H}^1(T(s)) - \mathcal{H}^1(K \cap D(s)).$$

We shall evaluate in more and more accurate ways this "excision energy". We start with the case where K has zero measure inside the disk $D(R)$.

3. The excision lemmas.

Excision Lemma 14.12. *Let K be a closed 1-set satisfying the property (M). If, for some s, $\partial D(s)$ does not meet K and if for every $\varepsilon > 0$ there is a small oscillation covering D_i (made of disks) of $K \cap D(s)$ with fixed constants μ and ν which do not change with ε such that*

$$\sum_i p_i \leq \varepsilon,$$

then

$$E(u_s, K(s)) - E(u, K) \leq 0.$$

Let us now consider the case where we simply know that the measure of K inside $D(R)$ is "small enough".

Excision Lemma 14.13. *Let K be a closed 1-set satisfying the minimality property (M). Assume that there is a small oscillation covering by disks $D_i = B(x_i, s_i)$, $i \in I$, of $K \cap D(R)$, with constants μ and ν and which satisfies*

$$(14.14) \qquad \forall s \in [\frac{R}{2}, R], \qquad \sum_{D_i \subset D(s)} p_i \leq \mu \mathcal{H}^1(K \cap D(s))$$

(which clearly is an intermediate case between (14.9) and (14.10)). Then there exists a constant $\gamma(\mu, \nu)$ such that if $\sum_{i \in I} p_i < \gamma R$, then

$$E(u_s, K(s)) - E(u, K) \leq 0$$

for some $s \in [\frac{R}{2}, R]$.

Both preceding elimination lemmas essentially use the fact that the covering is sparse. Unless its proof will be quite close in spirit, the next lemmas introduce a new parameter: if the small oscillation covering is "thin enough", namely if it is made of sets which are much smaller than $D(R)$, then an estimate can be again proved which entails the decrease of Mumford-Shah energy by excision. Let us first give a general lemma for estimating the excision energy.

Excision Energy Lemma 14.15. *Assume that $(D_i)_i$ is a small oscillation covering of $K \cap D(R)$ with constants μ and ν. Then there exists $s \in [\frac{R}{2}, R]$ and a constant $C(\mu, \nu)$, only depending on μ and ν, such that*

$$E(u_s, K(s)) - E(u, K) \leq C(\mu, \nu) \sup_i (\frac{p_i}{R})^{\frac{1}{2}} \sum_i p_i - \mathcal{H}^1(K \cap D(\frac{R}{2})).$$

We now state several immediate consequences of the Excision Energy Lemma.

Corollary 14.16. *The constants μ and ν being given, assume that a small oscillation covering of $K \cap D(R)$ satisfies $\sup_i p_i \leq \alpha R$ for a constant α small enough. Then $E(u_s, K(s)) - E(u, K) \leq 0$.*

In other terms, whenever all the sets of a small oscillation covering have a small enough perimeter with respect to R, we may excise a subdisk $D(s)$ with $s \geq \frac{R}{2}$ without increasing the Mumford-Shah energy. In the next chapter, we shall deduce from this simple fact several geometric properties of the minimal segmentations. As an immediate consequence of the previous corollary we obtain

Excision Lemma 14.17. *Assume that a given small oscillation covering D_i satisfies $\sup_i p_i \leq \alpha R$ for a constant α small enough with respect to μ, ν. Then if K satisfies the minimality assumption (M) one has $\mathcal{H}^1(K \cap D(\frac{R}{2})) = 0$.*

We shall say that a point of a 1-set K is *almost isolated* if it has a neighborhood whose intersection with K has zero $\mathcal{H}^1$ measure. We say that a closed 1-set K is *essential* if it has no almost isolated points. We shall see in the next chapter (by using the first Excision Lemma 14.12) how a minimal set can always be assumed to be essential by eliminating its almost isolated points. After we assume such a property, the above Elimination Lemma can be trivially reformulated in the following way.

Elimination Theorem 14.18. *Assume that a given small oscillation covering D_i satisfies $\sup_i p_i \leq \alpha R$ for a constant α small enough with respect to μ, ν. Then if K satisfies the minimality assumption (M) and is an essential set, one has $K \cap D(\frac{R}{2}) = \emptyset$.*

The remainder of this chapter is devoted to the proof of the excision lemmas and of some intermediate steps. However, if the reader wants to see better the sense of these abstract results and wishes to understand how the small oscillation coverings which have just been introduced naturally appear in several applications, he can pass to the next chapter and see the theorems therein. Nothing more than the knowledge of the statement of the excision lemmas is needed before reading those results.

4. The Interpolation Lemma.

The first point that we have to check is the problem of defining a good excision of u when $T(s) \neq \emptyset$. The difficulty is that when we excise $D(s)$ from K, we have to redefine u. We replace u by a function u_s such that $u_s = u$ outside D, and we wish that the restriction v_s of u_s to $D(s)$ is a smooth extension of u inside $D(s)$. Now, whenever K meets $\partial D(s)$, u may have jumps on $\partial D(s)$ and we cannot just set $v_s = u$ on $\partial D(s)$:

because of the jumps, such a function could not be easily extended to a function belonging to $H^1(D(s))$. So we shall add to K some arcs contained in $\partial D(s)$ which contain $K \cap D(s)$, and define v_s on these arcs as an interpolation of u. For this, we have to "pay" from the energy viewpoint and the price is of course the length of the union $T(s)$ of the arcs of $D(s)$ thus added to the segmentation K. Note, however, that this technique (developed in the next lemma) is not needed when $\partial D(s)$ does not meet K, which is the case for the first two "elimination lemmas" stated above.

As a first step, we have to know a bound on the number of the arcs which will be required for $T(s)$ and of the regularity of u on the remaining part of $D(s)$. Such bounds will be established for s in a large enough subset of $[\frac{R}{2}, R]$ in the next lemma, based on a straightforward use of Fubini's Theorem.

We denote by $J(s)$ the set of the indices i such that $D_i \cap \partial D(s) \neq \emptyset$ and by Card $J(s)$ its cardinality.

Fubini Lemma 14.19. *Let K satisfy the minimality property (M) and $(D_i)_{i\in I}$ be a covering satisfying (14.9). Then there is a subset $F \subset [\frac{R}{2}, R]$, with measure larger than $\frac{R}{4}$, such that for any s in F and some universal constant C,*

(14.20) $$Card(J(s)) \leq C, \quad and$$

(14.21) $$\int_{\partial D(s)} |\nabla u|^2 \leq C.$$

In addition, we can assume that for any s in F, $D(s)$ is reliable with respect to K.

Proof. We showed in the preceding chapter (Definition (13.12)) that $D(s)$ is reliable for almost every s, so that the last announced property is satisfied. It is easily proved (see Lemma (14.43) at the end of this chapter) that

$$\int_0^R (\mathrm{Card}J(s))ds \leq \sum_i p_i.$$

Since K has the minimality property (M) (and therefore (14.5)), we deduce (14.20) and (14.21) from (14.9) and (14.5) using Fubini's Theorem. The constants C appearing in (14.20) and (14.21) are obtained by multiplying the constants in (14.9) and (14.5) by 8. $\square$

We are in a position to choose a convenient set $T(s)$ and a conveniently smooth extension of u on $T(s)$. Our aim is to define v_s starting from a suitably regular boundary value on $D(s)$.

We denote by $\omega_j^s(u)$ the oscillation of a function u defined on $\partial(D_j \cap D(s))$.

Lemma 14.22 (How to excise). *Let K be a closed 1-set satisfying (M) and $(D_i)_i$ be a small oscillation covering of $K \cap D(R)$. Then there exists for any $s \in F$ a function $v_s \in C^1(\partial D(s))$ and a subset $T(s) \subset \partial D(s)$ satisfying the excision conditions of (14.1) and such that*

$$(14.23) \qquad \mathcal{H}^1(T(s)) \leq \frac{\pi}{2} \sum_{j \in J(s)} p_j,$$

$$(14.24) \qquad \int_{\partial D(s)} (\frac{\partial v_s}{\partial \tau})^2 \leq C(\nu).$$

Set $\tilde{u}_s = u$ on $B \cap D(s)$ and $\tilde{u}_s = v_s$ on $\partial D(s)$. Then there also exists a constant $C(\nu)$, only depending on ν, such that

$$(14.25) \qquad \omega_j^s(\tilde{u}_s) \leq C(\nu) \sum_{j \in J(s)} p_j^{\frac{1}{2}}.$$

Proof. When $\mathrm{Card}(K \cap \partial D(s)) = 0$, we take $T(s) = \emptyset$ and $v_s = u$. In the case where K meets $\partial D(s)$ at a single point z, we could choose a D_j containing z and define $T(s)$ as an arc of $\partial D(s)$ centered at z and with length $\frac{\pi}{2} p_j$. In order to deal with the general case, we prefer to define $T(s)$ as a union of arcs $c_k = (x_k, y_k)$ of $\partial D(s)$. Denote by J_k the set of indices j such that D_j meets c_k ; we impose that

$$(14.26) \quad \mathcal{H}^1(c_k) = min(2\pi s, \frac{\pi}{2} \sum_{j \in J_k} p_j) \quad \text{and} \quad c_k \supset (\partial D) \cap \bigcup_{j \in J_k} D_j.$$

This is easily achieved by induction on $\mathrm{Card}(J_k)$. We then set $v_s = u$ on $\partial D(s) \setminus T(s)$ and make a C^1 interpolation of u on the arcs c_k. Clearly, the Lipschitz constant L_k of v_s on c_k can be bounded by $2\left|\frac{u(x_k)-u(y_k)}{\mathcal{H}^1(c_k)}\right|$. By an obvious connectivity argument (K meets every ∂D_j at most one point), we have by (14.11), (14.21) and (14.26)

$$|u(x_k) - u(y_k)| \leq \sum_{j \in J_k} \omega_j(u) + \int_{c_k} |\nabla u| \leq \nu \sum_{j \in J_k} p_j^{\frac{1}{2}} + C(\sum_{j \in J_k} p_j)^{\frac{1}{2}}.$$

Since $\mathrm{Card} J_k \leq \mathrm{Card} J(s) \leq C$, we deduce that, for some new constant $C(\nu)$ only depending on ν,

$$L_k \leq C(\nu)(\sum_{j \in J(s)} p_j)^{-\frac{1}{2}} = C\mathcal{H}^1(c_k)^{-\frac{1}{2}}.$$

Therefore we can chose v_s so that

$$\int_{\partial D(s)} (\frac{\partial v_s}{\partial \tau})^2 \leq \sum_k C(\nu)\mathcal{H}^1(c_k)\mathcal{H}^1(c_k)^{-1} + \int_{\partial D \backslash T(s)} |\nabla u|^2 \leq C(\nu),$$

where $C(\nu)$ only depends on ν. The remaining statements follow by simple computations and are left to the reader. Let us show that the set $T(s)$ satisfies the connectedness condition of Definition 14.1. We claim that, if x and y are two points in the same connected component of K, we can find two indices i and j such that x and y are in the same connected component of $K \cap (\overline{D}_i \cup \overline{D}_j)$. Since the arc of $T(s)$ which contains x covers by construction a $\frac{\pi}{2}p_i$-neighbourhood of x and the arc which contains y a $\frac{\pi}{2}p_j$-neighbourhood of y, the two point must belong to the same arc of $T(s)$.

Let us finally prove the claim about the connected components of K. It follows from the property that every set D_i can have at most one point of K on its boundary. A first obvious consequence of such a property is that the intersection of every connected component of K with every set $\overline{D}_i$ is connected. Then fix x in K and let Y_x be the set of the points y which satisfy this property and K_x be the connected component of K which contains x. Of course we have $Y_x \subset K_x$. We are going to show that Y_x must be closed and open in K_x, so that by the connectivity of K_x the two sets are actually equal. In order to see that Y_x is open, fix y in Y_x and fix i and j in such a way that x and y are in the same connected component of $K \cap (\overline{D}_i \cup \overline{D}_j)$. If y is in the interior of one of both sets D_i and D_j, then the trace of K_x on the same set is contained in Y_x, since it is connected, so y is in the interior of Y_x. If y belongs to ∂D_j then we must have $y \in \overline{D}_i$ or $x \in \overline{D}_j$ because x and y cannot otherwise be connected in $K \cap (\overline{D}_i \cup \overline{D}_j)$, since no other point of K can be on ∂D_j. In any case, by exchanging if necessary the roles of i and j, we can assume that x and y belong to the same $\overline{D}_i$. Then we can find a new index k such that y belongs to D_k and, since x and y already are connected by K on $\overline{D}_i$, we are back to the previous case. The closedness of Y_x can be checked in a simple way. Let $(y_n)_{n \in \mathbb{N}}$ be a sequence of points in Y_x. Taking advantage of the finiteness of the set of the indices, we can fix two indices i and j which do not depend on n in such a way that x and y_n are in the same connected component of $K \cap (\overline{D}_i \cup \overline{D}_j)$. If $(y_n)_{n \in \mathbb{N}}$ converges to a point y, y must belong to Y_x since the connected components of K are closed sets.

In conclusion, we have shown that the property enjoyed by the points in Y_x holds for every y in K_x. So if $y \in K_x$ we have in particular $d(x,y) \leq \frac{1}{2}(p_i + p_j)$ for two indexes i and j such that $x \in D_i$ and $y \in D_j$.
$\square$

5. Energy jump estimates.

As the preceding one, the next lemma is simple when the boundary of the "excised disk" $D(s)$ does not meet K. So, we divide its conclusions in two items, although we shall directly prove the second which is concerned with the general case and which includes the first one (modulo a constant). We denote by $I(s)$ the set of indices i such that $D_i \subset D(s)$ and by $J(s)$ the set of indices such that D_i meets $\partial D(s)$.

Lemma 14.27. *Let $s \in F$, $T(s)$, and $v_s \in C^1(\partial D(s))$ be as in Lemmas 14.19 and 14.22. Extend v_s inside $D(s)$ as the solution of $-\Delta v_s + v_s = g$. Then the "excision jump of energy", $I_{D(s)}(v_s) - I_{D(s)}(u)$, satisfies for almost every s in F,*

(i) if no D_i meets $\partial D(s)$:

$$I_{D(s)}(v_s) - I_{D(s)}(u) \le 4 \sum_{i \in I(s)} \left(\int_{\partial D_i} |\nabla u| \right) \left(\int_{\partial D_i} |\nabla v_s| \right) + 2\mathcal{L}_2(B \cap D(s)).$$

(ii) if $J(s)$ is not empty:

$$\begin{aligned}
I_{D(s)}(v_s) - I_{D(s)}(u) \le\ & 4 \sum_{i \in I(s) \setminus J(s)} \left(\int_{\partial D_i} |\nabla u| \right) \left(\int_{\partial D_i} |\nabla v_s| \right) \\
& + \sum_{j \in J(s)} (\omega_j^s(\tilde{u}) + \omega_j^s(v_s)) \int_{\partial(D_j \cap D(s))} (|\nabla u| + |\nabla v_s|) \\
& + 5\mathcal{L}_2(B \cap D(s)).
\end{aligned}$$

Proof. We shall omit in this proof the indices s and write v, D, ω_j, etc. instead of v_s, $D(s)$, ω_j^s...Although ω_j was already used to denote the oscillation on ∂D_j, this will not introduce any ambiguity. Indeed, one has $\omega_j = \omega_j^s$ whenever D_j does not meet $\partial D(s)$. The proof essentially is an application of Green Formula Lemmas 13.10, 13.13 and 13.15 proved in the previous chapter. Setting $B = \bigcup_i D_i$, we can write

$$I_D(v) - I_D(u) = I_{D \setminus B}(v) - I_{D \setminus B}(u) + I_{D \cap B}(v) - I_{D \cap B}(u).$$

For almost every s in F, $\partial D(s)$ meets the ∂D_i at a finite set of points. Thus we can assume that $D \setminus B$ is a piecewise C^1 domain and we can apply the Green Lemma 13.10 which yields, noting that by Lemma 14.22, $u = v$ on $\partial D \setminus B$, $\tilde{u} = v$ on $\partial D \cap B$ and writing

$$\partial(D \setminus B) = (\partial D \setminus B) \cup (\partial B \cap D)) = \partial(B \cap D) \setminus (\partial D \setminus B) \cup \partial D \cap B,$$

$$\text{(14.28)} \quad \begin{aligned} I_{D\setminus B}(v) - I_{D\setminus B}(u) &= -\int_{\partial(D\setminus B)} (v-u)(\frac{\partial u}{\partial n} + \frac{\partial v}{\partial n}) \\ &= -\int_{\partial(D\cap B)} (v-\tilde{u})(\frac{\partial u}{\partial n} + \frac{\partial v}{\partial n}). \end{aligned}$$

Indeed (see Lemma 14.22), $\tilde{u} = v$ on ∂D by definition and, by the orientation rule adopted for the Green Formula, $\frac{\partial u}{\partial n}$ and $\frac{\partial v}{\partial n}$ change sign when computed on ∂B instead of $\partial(D\setminus B)$. The relation (14.28) yields

$$\text{(14.29)} \quad \begin{aligned} I_{D\setminus B}(v) - I_{D\setminus B}(u) &= -\int_{\partial(D\cap B)} v\frac{\partial u}{\partial n} + \int_{\partial(D\cap B)} \tilde{u}\frac{\partial v}{\partial n} \\ &\quad - \int_{\partial(D\cap B)} v\frac{\partial v}{\partial n} + \int_{\partial(D\cap B)} \tilde{u}\frac{\partial u}{\partial n} \\ &= I_1 + I_2 + I_3 + I_4. \end{aligned}$$

We now proceed to estimate the four preceding integrals. By Localization of Green Formula Lemma (13.15),

$$\begin{aligned} I_1 + I_2 &\leq 2 \sum_{i\in I\setminus J} (\omega_i(v)\int_{\partial D_i} |\nabla u| + \omega_i(u)\int_{\partial D_i} |\nabla v|) \\ &\quad + 2\sum_{j\in J}(\omega_j(v) + \omega_j(\tilde{u}))\int_{\partial(D_j\cap D_s)} (|\nabla u| + |\nabla v|) \\ &\quad + 2\mathcal{L}_2(B\cap D) \\ &\leq 4 \sum_{i\in I\setminus J} ((\int_{\partial D_i} |\nabla u|)(\int_{\partial D_i} |\nabla v|)) \\ &\quad + 2\sum_{j\in J}(\omega_j(v) + \omega_j(\tilde{u}))\int_{\partial(D_j\cap D_s)} (|\nabla u| + |\nabla v|) \\ &\quad + 2\mathcal{L}_2(B\cap D). \end{aligned}$$

By Green Formula 13.13, and since $|v|, |\Delta v|, |v-g|$ are bounded by 1,

$$\text{(14.30)} \quad \begin{aligned} I_3 + I_{B\cap D}(v) &= -\int_{\partial(B\cap D)} v\frac{\partial v}{\partial n} + \int_{B\cap D} (|\nabla v|^2 + (v-g)^2) \\ &\int_{B\cap D} (-\Delta v + (v-g)^2) \leq 2\mathcal{L}_2(B\cap D). \end{aligned}$$

In order to evaluate the last term, we set $B = B' \cup B''$, where $B' = \bigcup_{j\in J}(D_j \cap D)$ and $B'' = B \setminus B'$. We can split I_4 as $I_5 + I_6$, where

$$\text{(14.31)} \quad I_5 = \int_{\partial B'} \tilde{u}\frac{\partial u}{\partial n} \leq \sum_{j\in J}\omega_j(\tilde{u})\int_{\partial(D_j\cap D_s)} |\nabla u|,$$

and since, on B'', $u = \tilde{u}$,

$$I_6 - I_{B \cap D}(u) = \int_{\partial B''} u \frac{\partial u}{\partial n} - \int_{B \cap D} (|\nabla u|^2 + (u - g)^2)$$

$$= \int_{B''} (|\nabla u|^2 + u \Delta u) - \int_{B \cap D} (|\nabla u|^2 + (u - g)^2)$$

$$= \int_{B''} (|\nabla u|^2 + u \Delta u) - \int_{B''} (|\nabla u|^2 + (u - g)^2),$$

since $B'' \subset B \cap D$. Using the fact that $|u|, |\Delta u|$ are less than 1, we obtain

(14.32) $I_6 - I_{B \cap D}(u) \leq \mathcal{L}_2(B'') \leq \mathcal{L}_2(B \cap D).$

Putting together the relations (14.28) to (14.32) achieves the proof. $\square$

6. Proof of the excision lemmas.

We now pass to the proof of the key results of this chapter, starting with the second excision lemma; the second will be obtained as an easier variant.

Proof of the Excision Lemma 14.13. Fix $\gamma > 0$, $\lambda > 1$ and consider the balls $D_i' = B(x_i, \lambda s_i)$. When $\lambda \gamma \leq \frac{1}{4}$, we deduce from the relation $\sum_{i \in I} p_i < \gamma R$ that

$$\sum_{i \in I} \mathcal{H}^1(\partial D_i') \leq \lambda \sum_{i \in I} \mathcal{H}^1(\partial D_i) \leq \lambda \gamma R \leq \frac{R}{4}.$$

By using the same obvious argument as in Fubini Lemma 14.19, we can deduce the existence of a positive number $s \in [\frac{R}{2}, R]$ such that

(14.33) $\partial D(s) \cap \bigcup_{i \in I} D_i' = \emptyset$

and

(14.34) $\int_{\partial D(s)} |\nabla u|^2 d\mathcal{H}^1 \leq 12\pi.$

As a consequence of the relation (14.33), the disks D_i contained in $D(s)$ are a covering of $D(s) \cap K$. Let us consider the set $K(s)$ and the function u_s obtained by excision of the disk $D(s)$ from K and u, according to Definition 14.1. Note that, by (14.33), the two definitions have to be considered in a particularly simple case and don't require the use of

Interpolation Lemma 14.22. For the same reason, only the simpler item (i) of Lemma 14.27 is required in order to estimate the energy jump $I(K(s)) - I(K)$. Denoting by $I(s)$ the set of the indices i such that $D_i \subset D(s)$, we obtain by Excision Lemma 14.27,

$$
(14.35) \quad
\begin{aligned}
I(K(s)) - I(K) \leq & 4 \sum_{i \in I(s)} \left(\int_{\partial D_i} |\nabla u| d\mathcal{H}^1 \right) \left(\int_{\partial D_i} |\nabla v_s| d\mathcal{H}^1 \right) \\
& + 2\mathcal{L}_2(B \cap D).
\end{aligned}
$$

The first integral in the above sum is directly estimated by

$$
(14.36) \quad \int_{\partial D_i} |\nabla u| d\mathcal{H}^1 \leq C s_i^{\frac{1}{2}}
$$

because (D_i) is a small oscillation covering. For the second one, we are in a position to apply the gradient estimate (13.27) proved in Chapter 13. Since we know by (14.33) that for every index i in $I(s)$ the set $D_i' = \lambda D_i$ is contained in $D(s)$, we deduce that

$$
\forall x \in D_i, \quad d(x, \partial D(s)) \geq (\lambda - 1)s_i
$$

and therefore that, for an absolute constant C deduced from (14.34) and Lemma 13.27,

$$
(14.37) \quad \forall i \in I(s) \int_{\partial D_i} |\nabla v_s| \leq \frac{C s_i^{\frac{1}{2}}}{\sqrt{(\lambda - 1)}}.
$$

We finally remark that

$$
\mathcal{L}_2(B \cap D) \leq C \sum_{i \in I(s)} s_i^2 \leq C \gamma R \sum_{i \in I(s)} s_i.
$$

Using this last inequality and combining (14.35) with (14.36), (14.37) and (14.14), we get for various absolute constants, both denoted by C,

$$
(14.38) \quad
\begin{aligned}
I(K(s)) - I(K) & \leq C \left(\frac{1}{\sqrt{(\lambda - 1)}} + \gamma \right) \sum_{i \in I(s)} s_i \\
& \leq C \left(\frac{1}{\sqrt{(\lambda - 1)}} + \gamma \right) \mathcal{H}^1(K \cap D(s)).
\end{aligned}
$$

Thus, if we take λ large enough to ensure $C\frac{1}{\sqrt{(\lambda-1)}} + \gamma \le 1$, as we can do when γ is suitably small, we obtain

$$E(u_s, K(s)) - E(u, K) \le I(K(s)) - I(K) - H^1(K \cap D(s)) \le 0.$$

□

Proof of the Excision Lemma 14.12. We do not need to insist on this proof because it is a simplification of the preceding one. Here again we excise $D(s)$ from K with $T(s) = \emptyset$. Note that s is fixed and has not to be selected. In order to fix λ, we make sure as above that the disks $D_i' = \lambda D_i$ do not meet $\partial D(R)$. This is achieved by taking $\varepsilon < \frac{d(K, \partial D(s))}{\lambda}$, which is the only condition imposed on λ. So the above obtained relation

$$(14.38) \qquad I(K(s)) - I(K) \le C\left(\frac{1}{\sqrt{(\lambda - 1)}} + \gamma\right) \sum_{i \in I(s)} s_i$$

still holds, and since we can make $\sum_{i \in I(s)} s_i$ arbitrarily small, we obtain

$$E(u_s, K(s)) - E(u, K) \le I(K(s)) - I(K) \le 0.$$

□

The next proof uses the same arguments as both preceding ones but with an additional technical difficulty which was treated in Lemma 14.22. When the perimeter $\sum_i p_i$ of the covering D_i is large, we cannot ensure that the boundary of the excision disk, $\partial D(s)$, does not meet some ∂D_i. For that reason, the excision energy may sometimes be infinite. Now, we shall prove that its mean value when s varies between R and $\frac{R}{2}$ can be made very small when the covering is "thin".

Proof of the Excision Lemma 14.15. We define v_s and $T(s)$ as explained in Lemma 14.22 and proceed to estimate all terms arising in the inequality obtained in Lemma 14.27(ii). By Lemma 14.22, we know that v_s satisfies

$$(14.24) \qquad \int_{\partial D(s)} (\frac{\partial v_s}{\partial \tau})^2 \le C(\nu),$$

and the constants only depend on ν. Therefore, we can apply Lemma 13.28 so that the gradient of v_s inside D_s is estimated in terms of the distance from the boundary by

$$(14.39) \qquad \forall x \in D(s), \quad \forall s \in F, \quad |\nabla v_s(x)| \le \frac{C(\nu)}{(s - |x|)^{\frac{1}{2}}}.$$

It is easily deduced from (14.24) and (14.39) that

$$(14.40) \qquad \forall i \in J(s), \quad \omega_i^s(v_s) \leq C p_i^{\frac{1}{2}}.$$

Indeed, let x and y be two points of $\partial(D_i \cap D(s))$. We consider the closest points $\tilde{x}$ and $\tilde{y}$ of $\partial D(s)$, so that the lines $x\tilde{x}$ and $y\tilde{y}$ contain the center of $D(s)$. Integrating,

$$|v_s(x) - v_s(y)| \leq \int_{[x,\tilde{x}] \cup (\tilde{x},\tilde{y}) \cup [\tilde{y},y]} |\nabla v_s(x)| d\mathcal{H}^1(x),$$

where $(\tilde{x}, \tilde{y})$ denotes the shortest arc of $D(s)$ connecting $\tilde{x}$ and $\tilde{y}$, and using (14.39), (14.24) yields (14.40). For the commodity of the reader, we recall the basic estimates satisfied in F by Fubini Lemma 14.19,

$$(14.20) \qquad \text{Card}(J(s)) \leq C,$$

$$(14.21) \qquad \int_{\partial D(s)} |\nabla u|^2 \leq C,$$

and by Lemma 14.22

$$(14.25) \qquad \omega_j^s(\tilde{u}) \leq C(\nu) \sum_{j \in J(s)} p_j^{\frac{1}{2}}.$$

Using the three preceding inequalities, (14.11), (14.8) and the Cauchy-Schwarz inequality, we obtain for various constants C only depending on μ and ν:

$$\forall i \in I(s) \setminus J(s), \quad \int_{\partial D_i} |\nabla u| \leq C p_i^{\frac{1}{2}},$$

$$(14.41) \qquad \forall i \in J(s), \quad \omega_i^s(\tilde{u}) \leq C \sum_{j \in J(s)} p_j^{\frac{1}{2}} \leq C R^{\frac{1}{2}},$$

$$(\sum_{j \in J(s)} p_j^{\frac{1}{2}})^2 \leq C \sum_{j \in J(s)} p_j.$$

By Lemma 14.27(ii), we know that the jump of the Mumford-Shah energy by excision of D_s from K satisfies

$$E(K(s)) - E(K) \leq I_{D(s)}(v_s) - I_{D(s)}(u) + \mathcal{H}^1(T(s)) - \mathcal{H}^1(K \cap D(s))$$

$$\leq 4 \sum_{i \in I \setminus J(s)} ((\int_{\partial D_i} |\nabla u|)(\int_{\partial D_i} |\nabla v_s|)) +$$

$$2 \sum_{j \in J(s)} (\omega_j(\tilde{u}_s) + \omega_j(v_s)) \int_{\partial(D_j \cap D_s)} (|\nabla u| + |\nabla v_s|) +$$

$$5 \mathcal{L}_2(B \cap D(s)) + C \sum_{j \in J(s)} p_j - \mathcal{H}^1(K \cap D(\frac{R}{2})),$$

where we have used the first relation (14.23) of Lemma 14.22, $\mathcal{H}^1(T(s)) \leq C\sum_{j\in J(s)} p_j$. Using (14.21), (14.25), (14.39), (14.40) and (14.41), we obtain for constants C only depending on μ and ν,

$$E(K(s)) - E(K) \leq C\sum_{i\in I} p_i^{\frac{1}{2}} \int_{\partial D_i} \frac{d\mathcal{H}^1(x)}{(s-|x|)^{\frac{1}{2}}} + C\sum_{j\in J(s)} p_j$$

$$+ CR^{\frac{1}{2}} \sum_{j\in J(s)} \int_{D(s)\cap\partial D_j} \frac{d\mathcal{H}^1(x)}{(s-|x|)^{\frac{1}{2}}}$$

$$+ 5\mathcal{L}_2(B\cap D(s)) - \mathcal{H}^1(K\cap D(\frac{R}{2})).$$

We now make a mean value estimate on $s \in F$ because the right-hand integrals of the preceding inequality can be large and even infinite for some "bad" values of s, for instance, when some ∂D_j is contained in and tangent to $D(s)$. The mean value on F, however, is easily controlled in terms of the p_i. Indeed, using Lemma 14.43 (below) we obtain

$$\frac{1}{\mathcal{H}^1(F)} \int_F (E(K(s)) - E(K))ds \leq$$

$$CR^{-1}(\sum_{i\in I} p_i^2 + R^{\frac{1}{2}}\sum_i p_i^{\frac{3}{2}}) - \mathcal{H}^1(K\cap D(\frac{R}{2})).$$

Since $\mathcal{H}^1(F) \geq \frac{R}{4}$, the same inequality (with a larger constant C) holds with the left-hand side replaced by $E(K(s)) - E(K)$ for some s in $[\frac{R}{2}, R]$. Since we know that $\sum_i p_i \leq CR$, we get for such an s

$$E(K(s)) - E(K) \leq$$

(14.42)

$$\leq C\left(\frac{\max(p_i)}{R}\right)^{\frac{1}{2}} \sum_i p_i - \mathcal{H}^1(K\cap D(\frac{R}{2})). \quad \square$$

Lemma 14.43. *Let $D(s)$, $0 \leq s \leq R$, $(D_i)_{i\in I}$, p_i, $J(s)$ be as above. Then*

(14.44)

$$\int_0^R \left(\int_{\partial D_i} \frac{d\mathcal{H}^1(x)}{((s-|x|)^+)^{\frac{1}{2}}}\right)ds \leq 2p_i^{\frac{3}{2}},$$

(14.45)

$$\int_0^R \left(\sum_{j\in J(s)} \int_{D(s)\cap\partial D_j} \frac{d\mathcal{H}^1(x)}{(s-|x|)^{\frac{1}{2}}}\right)ds \leq 2\sum_{i\in I} p_i^{\frac{3}{2}},$$

(14.46)
$$\int_0^R (\sum_{j \in J(s)} p_j) ds \le \sum_{i \in I} p_i^2,$$

(14.47)
$$\int_0^R CardJ(s) ds \le \sum_{i \in I} p_i.$$

Proof. Set $s_i = \inf \{s, D(s) \cap D_i \ne \emptyset\}$. Then we have, for any $x \in D_i$, $s_i \le |x| \le s_i + p_i$, and by Fubini's Theorem,

$$\int_0^R (\int_{\partial D_i} \frac{d\mathcal{H}^1(x)}{((s - |x|)^+)^{\frac{1}{2}}}) ds \le \int_{\partial D_i} (\int_{s_i}^{s_i + p_i} \frac{ds}{((s - |x|)^+)^{\frac{1}{2}}}) d\mathcal{H}^1(x) \le 2p_i^{\frac{3}{2}}.$$

In order to prove (14.45), we notice that $\{(x, i, s), x \in \partial D_i, i \in J(s)\} \subset \{(x, i, s), s_i \le s \le s_i + p_i\}$. Therefore,

$$\int_0^R (\sum_{j \in J(s)} \int_{D(s) \cap \partial D_j} \frac{d\mathcal{H}^1(x)}{(s - |x|)^{\frac{1}{2}}}) ds$$

$$\le \sum_{i \in I} \int_{\partial D_i} (\int_{s_i}^{s_i + p_i} (\frac{ds}{((s - |x|)^+)^{\frac{1}{2}}})) d\mathcal{H}^1(x)$$

$$\le 2 \sum_{i \in I} p_i^{\frac{3}{2}}.$$

The proof of the other relations is analogous and simpler. $\quad\square$

Chapter 15

DENSITY PROPERTIES
AND EXISTENCE THEORY
FOR THE MUMFORD-SHAH MINIMIZERS

In this chapter, we shall find a first application of the elimination techniques which we have introduced in the preceding chapter. We shall prove more and more precise density properties for minimal segmentations and, more generally, for segmentations satisfying the minimality property (M). Recall that this property essentially states that no excision of a disk can decrease the Mumford-Shah energy of the segmentation. To make a long story short, let us say that we shall prove (in this order):

• *The Essentiality Property: Around every point x of K one has that $\mathcal{H}^1(K \cap D(R)) > 0$, where $D(R)$ denotes the disk $B(x, R)$. In other terms, no point of K is "isolated".*

• *The Uniform Density Property: There is a universal constant β such that at every point $x \in K$, $\mathcal{H}^1(K \cap D(R)) > \beta R$. This clearly is a stronger nonsparseness property for K.*

• *The Concentration Property: For every $\varepsilon > 0$, there is $\alpha > 0$ such that every disk $D(R)$ centered on the edge set K contains a subdisk D' with radius larger than αR inside which K is "concentrated", that is $\mathcal{H}^1(K \cap D') \geq (1 - \varepsilon)\mathrm{diam}(D')$. This property clearly implies the Uniform Density Property.*

• *The First Projection Property: There is a universal constant β_1 such that $\forall x \in K$, $\mathcal{H}^1(p_1(K \cap D(R))) + \mathcal{H}^1(p_2(K \cap D(R))) \geq \beta_1 R$, where p_1 and p_2 denote the projectors in two orthogonal directions.*

We shall also prove a Second Projection Property, which is in the same relation to the first as the Concentration Property is to the Uniform Density Property. We could have directly stated and proved the two last density properties, which are stronger, but this would have made the exposition tedious and technical. As a matter of fact, each one of the above-mentioned properties will be used in the proof of existence of minimizers for the Mumford-Shah energy. To be more precise, we shall make in the second part of this chapter the following deductions.

• *From the Uniform Projection Property we deduce, by using the results of Chapter 12 about projections of 1-sets, that the Mumford-Shah minima and quasiminima (in the sense of Property (M)) must be rectifiable.*

• *From the Uniform Projection Property, we deduce that any minimal or quasiminimal (in the sense of property (M)) rectifiable segmentation can be approximated by a finite set of curves.*

• *From both the Projection Property and the Uniform Concentration Property, we deduce, by using the results of Chapter 10, that the set of segmentations satisfying (M) is compact and that the minimum of the Mumford-Shah energy is attained for some uniformly rectifiable segmentation.*

We end the chapter by showing that the minimal segmentation may be nonunique, which somehow matches the computer scientist's intuition that more than one segmentation can be "good" for a given image.

1. The Essentiality Property.

In this chapter as in the preceding, we consider "segmentations", that is, closed 1-sets denoted by K of an image domain Ω. (For instance, Ω can be a rectangle.) We also consider the associated function $u = u_K$ which minimizes the energy $I(u, K)$ as defined in the beginning of Chapter 13. Recall (Chapter 14) that a segmentation K is *essential* if for every $x \in K$ and every $r > 0$ one has $\mathcal{H}^1(K \cap B(x, r)) > 0$. An essential segmentation has no isolated points.

Theorem 15.1. *Let K be a closed 1-set satisfying (M). Then there is an essential closed 1-set $\tilde{K} \subset K$ such that $E(K) = E(\tilde{K})$ and $\mathcal{H}^1(\tilde{K} \backslash K) = 0$. In addition, $u_{\tilde{K}} = u_K$ almost everywhere.*

In other terms, we can "eliminate" from K all its "almost isolated points" without changing the energy or the approximating function.

Proof. Let x be an almost isolated point of K. Recall that by "almost isolated point" we mean a point x such that $\mathcal{H}^1(K \cap B(x, R)) = 0$ for some R. Let us set as usual $D(s) = B(x, s)$. By the Fubini Lemma, for almost every $s \in [\frac{R}{2}, R]$, K does not meet the boundary of $D(s)$. For any $0 < \varepsilon < \frac{\mathrm{d}(K, \partial D(s))}{2}$, we can find a covering $(D_i) = (B(x_i, r_i))$ of $K \cap D(s)$ such that $\sum_i r_i < \varepsilon$. (This is a direct consequence of the definition of the Hausdorff measure.) Since K satisfies the property (M), we know that $u = u_K$ satisfies $\int_{B(x_i, r_i)} |\nabla u|^2 \leq C p_i$, where C denotes a universal constant. Therefore we can select $s_i \in [r_i, 2r_i]$ satisfying $\int_{\partial B(x_i, s_i)} |\nabla u|^2 \leq C$ such that $\partial B(x_i, s_i)$ does not meet K. Consequently, the disks $B(x_i, s_i)$ make a small oscillation covering of $K \cap D(s)$ satisfying $\sum_i s_i \leq 2\varepsilon$. Since this can be obtained for every ε, we can apply Excision Lemma 14.12, which asserts that $E(u_s, K(s)) \leq E(u, K)$. So we can excise the open disk $D(s)$ from K and therefore eliminate all isolated points of K belonging to $D(s)$. Since we can iterate this process for all disks $B(x, r)$ with rational radii and rational coordinates such that $\mathcal{H}^1(K \cap B(x, r)) = 0$, we obtain, after completing all excisions, a closed set $\tilde{K}$ which is essential and has the same measure as K. Let us now prove that $u_{\tilde{K}} = u_K$. Indeed, since u_K minimizes $I(u, K)$ and $\tilde{K} \subset K$, we have $I(u_{\tilde{K}}) \geq I(u)$. On the other side, we know from the preceding proof that $I(u_{\tilde{K}}) \leq I(u_K)$ and therefore $I(u_K) = I(u_{\tilde{K}})$. Since the functional I is strictly convex on $H^1(\Omega \setminus K)$, we conclude that $u_K = u_{\tilde{K}}$ almost everywhere. $\square$

As a consequence of the preceding result, from now on we may assume that the considered segmentations which satisfy the property (M) also are essential.

2. The Uniform Density Property.

Definition 15.2. *We say that a closed subset K of Ω satisfies the Uniform Density Property if there exists a constant β such that for every disk $D(R)$ contained in Ω with radius $R \leq 1$ and centered at a point of K,*

$$(15.3) \qquad\qquad \mathcal{H}^1(K \cap D(R)) \geq \beta R.$$

A trivially equivalent version of the above definition is the following.

Definition 15.4. *We say that a closed subset K of Ω satisfies the Uniform Density Property if there exists a constant β such that for every disk $D(R)$ contained in Ω with radius $R \leq 1$ such that*

$$\mathcal{H}^1(K \cap D(R)) \leq \beta R,$$

one has $K \cap D(\frac{R}{2}) = \emptyset$.

It is clear that, if K is as in Definition 15.4, then it enjoys the property also according to Definition 15.2. Conversely, when $K \cap D(\frac{R}{2}) \neq \emptyset$, $D(R)$ contains a disk with radius $\frac{R}{2}$ centered at a point of K. So the reverse implication also holds, provided the value of the constant β is divided by 2.

We are going to show that a minimum of the Mumford-Shah functional enjoys the Uniform Density Property with an absolute constant β which does not depend on K and Ω. The main idea is that whenever (15.3) fails on a disk $D(R)$, we can find a small oscillation covering of $K \cap D(R)$, and therefore $K \cap D(\frac{R}{2})$ is empty.

Theorem 15.5 (Uniform Density). *If a closed 1-subset K of Ω is essential and satisfies the excision property (M), then it also enjoys the Uniform Density Property in correspondence of a universal constant β.*

In order to prove this theorem, let us first admit the following covering lemma.

Covering Lemma 15.6. *Let K be a closed essential 1-set of Ω which satisfies the minimality property (M). Let $D(R)$ be a disk contained in Ω and with center x in K such that $\mathcal{H}^1(K \cap D(R)) \leq \beta R$. Then we can find a small oscillation covering $(D_i = B(x_i, s_i))_{i \in I}$ of $K \cap D(R(1 - 50\beta))$, with universal oscillation constant ν, and such that for every $\rho \in [\frac{R}{2}, R]$,*

$$(15.7) \qquad \sum_{D_i \subset D(\rho)} s_i \leq 51 \mathcal{H}^1(K \cap D(\rho)).$$

Proof of the Uniform Density Theorem 15.5. We assume that $H^1(K \cap D(R)) \leq \beta R$ for some constant β and some disk $D(R)$ contained in Ω and centered on K. Since K is essential, we can apply the preceding Covering Lemma and conclude that there is a small oscillation covering (D_i) of $K \cap D(R(1 - 10\beta))$ such that (15.7) holds. We set $R' = (1 - 50\beta)R$. By the Elimination Lemma 14.13, this implies that whenever $\sum_i s_i < \gamma R'$, where γ is a universal constant, then $E(K(s)) \leq E(K)$ for some $s \in [\frac{R'}{2}, R']$. In order to have $\sum_i s_i < \gamma R'$, it is enough by (15.7) to assume for example $\beta < \frac{\gamma}{100}$. Now, the property (M) and the fact that $E(K(s)) \leq E(K)$ imply $\mathcal{H}^1(K \cap D(\frac{R'}{2})) = 0$ and therefore contradict the essentiality assumption. Thus the Uniform Density Property is true with any $\beta < \frac{\gamma}{100}$. $\square$

Proof of the Covering Lemma 15.6. Let us first consider the set H of points of K where

$$\overline{d}_K^1(x) = \limsup_{r \to 0} \frac{\mathcal{H}^1(K \cap B(x, r))}{2r} < \frac{1}{8}.$$

For every x in H, let $r(x)$ be a positive real number such that

$$(15.8) \qquad \frac{\mathcal{H}^1(K \cap B(x, r))}{2r} < \frac{1}{8}$$

for every $r < r(x)$. By the relation (8.18) in Corollary 8.17, we know that $\mathcal{H}^1(H) = 0$, because the upper density of a 1-set is almost everywhere larger than $\frac{1}{2}$. Thus, by Lemma 6.31, for every $\varepsilon > 0$ we can find a covering of H with balls $D_i = B(x_i, r_i)$ centered on H such that

$$(15.9) \qquad\qquad \sum_i r_i \leq \frac{\varepsilon}{2}$$

and that for each one of these balls, $2r_i < r(x_i)$. Therefore, by (15.8)

$$(15.10) \qquad\qquad \mathcal{H}^1(K \cap B(x_i, 2r_i)) \leq \frac{r_i}{2}.$$

We shall now apply the same argument based on the Fubini Theorem which has already been used several times in the previous chapter. From (15.10) we see that for every i we can find a set $F_i \subset [r_i, 2r_i]$ with a measure larger or equal than $\frac{r_i}{2}$ such that

$$(15.11) \qquad\qquad \forall s \in F_i, \qquad \partial B(x_i, s) \cap K = \emptyset .$$

From the Fubini Theorem and the property (M), we also deduce that there exists at least one $s_i \in F_i$ such that (see Fubini Lemma 14.19)

$$(15.12) \qquad\qquad \int_{\partial B(x_i, s_i)} |\nabla u|^2 d\mathcal{H}^1 \leq 12\pi.$$

If for every i we replace D_i by $B(x_i, s_i)$, we obtain a small oscillation covering of H consisting of balls and such that, by (15.9),

$$(15.13) \qquad\qquad \sum_i s_i \leq \varepsilon.$$

We fix in the following $\varepsilon = \mathcal{H}^1(K \cap D(\frac{R}{2})) \leq \beta R$. Let us now consider the points x of $K \cap D((1 - 50\beta)R)$ satisfying

$$(15.14) \qquad \bar{d}_k^1(x) = \limsup_{r \to 0} \frac{\mathcal{H}^1(K \cap B(x, r))}{2r} \geq \frac{1}{8}$$

and define, for each x, $\bar{r}$ as the supremum of the real numbers r such that $B(x, r) \subset D(R)$ and

$$\frac{\mathcal{H}^1(K \cap B(x, r))}{2r} > \frac{1}{10}.$$

We have $\bar{r} \leq 5\beta R$, because (using $B(x, \bar{r}) \subset D(R)$)

$$\beta R \geq \mathcal{H}^1(K \cap B(x, r)) \geq \frac{r}{5}.$$

In addition, we know from (15.14) that $\bar{r} > 0$. Note that $B(x, 10\bar{r}) \subset D(R)$ and that, by the maximality of $\bar{r}$, we have

$$\mathcal{H}^1(K \cap B(x, 10\bar{r})) \leq \frac{1}{10} 20\bar{r} \leq 2\bar{r}$$

and

(15.15) $$\mathcal{H}^1(K \cap B(x, \bar{r})) \geq \frac{1}{5}\bar{r}.$$

Since from the last two inequalities

$$\mathcal{H}^1(K \cap (B(x, 10\bar{r}) \setminus B(x, 5\bar{r}))) \leq \frac{9}{5}\bar{r},$$

we can find a subset F of $[5\bar{r}, 10\bar{r}]$ with measure larger or equal than $5\bar{r} - \frac{9}{5}\bar{r} = \frac{16}{5}\bar{r} \geq 3\bar{r}$ such that $\partial B(x, s) \cap K = \emptyset$ for every s in F. Therefore we can choose s in F in such a way that

$$\int_{\partial B(x,s)} |\nabla u|^2 d\mathcal{H}^1 \leq 12\pi,$$

as we have done before. We deduce from (15.15) that

(15.16) $$\mathcal{H}^1(K \cap D(R) \cap B(x, \frac{s}{5})) \geq \mathcal{H}^1(K \cap B(x, \bar{r})) \geq \frac{1}{5}\bar{r} \geq \frac{1}{50}s.$$

By Covering Lemma 7.1, applied to the balls $B(x, \frac{s(x)}{5})$, we can select a subset X of $(K \cap D(R(1 - 50\beta))) \setminus H$ such that, if $s(x)$ is the value introduced above in dependence of x,

$$(K \cap D(R(1 - 50\beta))) \setminus H \subset \bigcup_{x \in X} B(x, s(x)),$$

$$\forall x, y \in X : B(x, \frac{s(x)}{5}) \cap B(y, \frac{s(y)}{5}) = \emptyset.$$

By (15.16), we clearly have for every $Y \subset X$

(15.17) $$\sum_{x \in Y} s(x) \leq 50 \sum_{x \in Y} \mathcal{H}^1(K \cap D(R) \cap B(x, \frac{s(x)}{5}))$$
$$\leq 50\mathcal{H}^1(K \cap D(R) \cap \bigcup_{x \in Y} B(x, s(x))).$$

Using the compactness of $K \cap D(R)$ we can choose a finite covering of $K \cap D(R(1 - 50\beta))$ consisting of balls $B(x_k, s_k)$ selected from the two sets of balls introduced above. Property (15.7) is then an obvious consequence of (15.13), (15.17) and the choice $\varepsilon = \mathcal{H}^1(K \cap D(\frac{R}{2}))$. $\square$

3. The Concentration Property.

Definition. *We say that a closed subset K of Ω satisfies the* Atomization Condition *(Condition (A)) on a disk $D(R)$ for two given positive constants α and ε if every disk D contained in $D(R)$ with* diam $D \geq \alpha R$ *satisfies $\mathcal{H}^1(K \cap D) < (1-\varepsilon)$diam D.*

Definition. *We say that a closed subset K of Ω satisfies the Uniform Concentration Property in Ω if for every $\varepsilon > 0$ there exists $\alpha = \alpha(\varepsilon)$ such that, if $D(R) = D(x_0, R)$ is any disk contained in Ω with $x_0 \in K$ and $0 < R \leq 1$, then there exists a disk $D = B(x, r)$ contained in $D(R)$ such that*

$$\text{diam}\, D \geq \alpha \text{diam}\, D(R),$$

$$\mathcal{H}^1(K \cap D) \geq (1-\varepsilon)\text{diam}\, D.$$

The above definitions are clearly connected. In fact, if a set K does not satisfy the Uniform Concentration Property, one can find a positive number $\varepsilon \geq 0$ such that, whatever the choice of the constant α is, the Atomization Condition (A) is satisfied on a suitable disk. We point out that the Uniform Concentration Property trivially implies the Uniform Density Property (with a value of β given by $2\alpha(\varepsilon)(1-\varepsilon)$), in correspondence of any fixed value of ε.

Theorem 15.18. *Let K be a closed essential 1-set of Ω which satisfies the minimality assumption (M). Then K satisfies the Uniform Concentration Property. Moreover, given $\varepsilon \geq 0$, the choice of the constant $\alpha(\varepsilon)$ does not depend on K.*

Proof. It is a direct application of Elimination Theorem 14.18. Let us first sketch the proof. Assume that ε is given and that, for a given α, the Atomization Condition (A) holds on some disk $D(R)$. Then we shall find a small oscillation covering of $K \cap D(R')$, where $R' = (1-\alpha)R$, with a constant μ proportional to $(\beta)^{-1}$ where β is the constant appearing in the Uniform Density Lemma and with a constant ν proportional to $\varepsilon^{\frac{-1}{2}}$. This small oscillation covering will consist of disks with a diameter smaller than $2\alpha R$. By the Elimination Theorem this will lead to conclude that the center of $D(R)$ cannot belong to K if the constant α is small enough according to μ and ν. We shall deduce that the Atomization Condition (A) cannot be fulfilled on any disk centered on K, and therefore the Concentration Property must hold.

Let us now give some details. We fix $x \in K \cap D(R')$. If (A) holds we can fix a positive number r such that

(15.19) $$\mathcal{H}^1(K \cap D(x, r)) \leq (1-\varepsilon)2r$$

and such that $r \leq \alpha R$. Setting

$$F = \{\rho \in]0,\ r[\ |\ \mathrm{Card}\,(K \cap \partial D(x, \rho)) \leq 1\},$$

we see from (15.19) that the measure of F has to be larger than εr. At this point, using the minimality property (M) and Fubini Theorem as we have already done for the proof of 14.21, we find a number $s \in (0, r)$ such that

$$(15.20) \qquad \int_{\partial B(x,s)} |\nabla u|^2 \leq \frac{3\pi}{\varepsilon}.$$

By the Cauchy-Schwarz Inequality, we deduce that the oscillation constant of u on $B(x, s)$ is less than $\nu = \frac{\sqrt{6}\pi}{\sqrt{\varepsilon}}$. As in the proof of the Uniform Density Property, we can use Covering Lemma 7.1 in order to cover $K \cap D_{R'}$ with a locally finite family of balls $(B(x_i, s_i))_{i \in I}$, satisfying (15.20) and such that

$$(15.21) \qquad \forall i, j \in I, i \neq j,\ B(x_i, \frac{s_i}{5}) \cap B(x_j, \frac{s_j}{5}) = \emptyset.$$

By the Uniform Density Property, for every index i we have

$$\mathcal{H}^1(K \cap B(x_i, \frac{s_i}{5})) \geq \frac{\beta}{5} s_i,$$

which yields, by (15.21),

$$\sum_{i \in I} s_i \leq \frac{5}{\beta} \sum_{i \in I} \mathcal{H}^1(K \cap B(x_i, \frac{s_i}{5})) \leq \frac{5}{\beta} \mathcal{H}^1(K \cap D(R)).$$

Since the Uniform Density Property also implies that

$$\mathcal{H}^1(K \cap D(R)) \leq 3\pi R \leq \frac{6\pi}{\varepsilon\beta} \mathcal{H}^1(K \cap D(\frac{R}{2})),$$

the small oscillation covering has a constant μ which is easily computed in terms of β as $\mu = \frac{10\pi}{\beta} \frac{6\pi}{\varepsilon\beta}$. So we have found a small oscillation covering with the properties required in the beginning of this proof, and this makes the argument complete. $\square$

4. The Projection Properties.

The aim of this section is the discussion of a variant of the last argument which proves that the Mumford-Shah minimal segmentations are *regular* in the Besicovitch sense, introduced in Chapter 8. Such a result will be obtained by using the characterization, proved in Chapter 12, of regular sets *via* their projections on lines. Following the notation introduced therein, we shall denote by θ a unitary vector of $\mathbb{R}^2$ and by p_θ the orthogonal projector in the direction θ.

Definition. *A closed 1-set K of a rectangle Ω is said to satisfy the First Projection Property if there exists $\beta_1 > 0$ such that, for every disk $D(R)$ centered at a point of K and contained in Ω with $R \leq 1$ and for every θ,*

$$(15.22) \qquad \mathcal{H}^1\left(p_\theta(K \cap D(R))\right) + \mathcal{H}^1\left(p_{\theta\perp}(K \cap D(R))\right) \geq \beta_1 R.$$

Analogously to what happened with the Uniform Density Property and the Concentration Property, we can consider the "concentrated" version of the Projection Property. To this aim, when D is a disk and θ is an unitary vector, we shall use the notation D_θ in order to denote the open square circumscribed by D and with one of its sides parallel to the direction θ.

Definition 15.23. *A closed 1-set K of Ω is said to satisfy the Second Projection Property if, for every $\varepsilon \geq 0$, there exists a positive constant $\alpha_1 = \alpha_1(\varepsilon)$ such that every disk $D(R)$ centered at a point of K and contained in Ω, with $R \leq 1$, contains a disk $D(r)$ with radius $r \geq \alpha_1 R$ satisfying, for every θ,*

$$(15.24) \quad \mathcal{H}^1\left(p_\theta(K \cap (D(r))_\theta)\right) + \mathcal{H}^1\left(p_{\theta\perp}(K \cap (D(r))_\theta)\right) \geq (1-\varepsilon)\frac{\sqrt{2}}{2}r.$$

Note that $\sqrt{2}r$ is the length of the side of the square $S = (D(r))_\theta$. This makes clear the analogy between the above definition and the Uniform Concentration Property. Of course, the second Projection Property implies the first one and can be related to an Atomization Condition.

Definition 15.25. *We say that a closed subset K of Ω satisfies the Atomization Condition (A') on a disk $D(R) = B(x_0, R)$ for two given positive constants α_1 and ε if for every disk $D(r) = B(x, r)$ contained in $D(R)$ with $2r = \operatorname{diam} D \geq 2\alpha_1 R$ one can find an unitary vector θ such that*

$$(15.26) \quad \begin{aligned} \mathcal{H}^1\left(p_\theta(K \cap (D(r))_\theta)\right) &+ \mathcal{H}^1\left(p_{\theta\perp}(K \cap (D(r))_\theta)\right) \\ &\leq (1-\varepsilon)\frac{\sqrt{2}}{2}r. \end{aligned}$$

The set K does not satisfy the Second Projection Property if and only if one can find a positive number $\varepsilon \geq 0$ such that, whatever the choice of the constant α_1 is, Condition (A') is satisfied on a suitable disk. Notice that the Uniform Density Property is clearly enjoyed by a set which satisfies the First Projection Property.

We shall directly prove that any minimum (or a quasiminimum) of the Mumford-Shah functional satisfies the Second Projection Property. The proof of the first one will therefore be achieved as a trivial corollary. As we have already pointed out, we just need to consider a minor variant of the argument in the proof of the Concentration Property. In fact, we shall only be careful in the construction of a suitable small oscillation covering, and this will allow us to conclude as above. Even the construction of the covering will be completely similar to the above constructions and based on a proper use of Fubini Theorem and Covering Lemma 7.1. The variant will consist in finding a covering consisting of squares rather than of disks in order to take advantage of the assumption (A').

Theorem 15.27. *Let K be a closed essential 1-set of Ω which satisfies the minimality assumption (M). Then K satisfies the Second Projection Property. Moreover, given $\varepsilon \geq 0$, the choice of the constant $\alpha_1(\varepsilon)$ does not depend on K, namely, the property holds with a universal function $\varepsilon \to \alpha_1(\varepsilon)$.*

Proof. Fix ε and assume that, for a given α_1, Condition (A') holds on some disk $D(R)$. Then we shall find a small oscillation covering of $K \cap D(R')$, where $R' = (1 - \alpha_1)R$, with a constant μ only depending on the constant β appearing in the Uniform Density Lemma, a constant ν proportional to $\varepsilon^{-\frac{1}{2}}$. This covering will consist of squares with a side length smaller than $\sqrt{2}\alpha_1 R$. By the Elimination Theorem 14.18, this will lead to conclude that the center of $D(R)$ cannot be on K if the constant α is small enough according to μ and ν.

Fix $x \in K \cap D(R')$. If (A') holds we can fix a positive number $r \leq \alpha R$ and an unitary vector θ such that, if we set $S = (D(R))_\theta$ and denote by $r' = \sqrt{2}r$ the side length of S, we have

$$(15.28) \qquad \mathcal{H}^1\left(p_\theta(K \cap S)\right) + \mathcal{H}^1\left(p_{\theta\perp}(K \cap S)\right) \leq \frac{(1 - \varepsilon)}{2}r'.$$

If we denote by S_ρ the square which has the same center as S, the sides in the same directions and side length ρ, and by F the set defined as

$$F = \{\rho \in]0, r'[, \quad K \cap \partial S_\rho = \emptyset\},$$

by (15.26), we can say that the measure of F is larger than εr. Then, by the Fubini Theorem and Property (M), we find a number $s \in (0, r')$ such that

$$(15.29) \qquad \int_{\partial S_s} |\nabla u|^2 \leq \frac{3\sqrt{2}\pi}{\varepsilon}.$$

By the Cauchy-Schwarz Inequality, we deduce that ∂S_s has an oscillation constant ν proportional to $\varepsilon^{\frac{1}{2}}$. By Covering Lemma 7.1 we can cover

$K \cap D(R')$ with a locally finite family of squares S_i such that, if $B(x_i, s_i)$ is the ball circumscribed by S_i, then (15.21) holds. Arguing as in the proof of the Concentration Property, we conclude from this fact that the covering has a constant μ which only depends on β. $\square$

5. Existence theory for the Mumford-Shah minimizing segmentations.

In the last sections we have seen that a minimum K of the Mumford-Shah functional satisfies some properties which have meaningful relations with the survey of Geometric Measure Theory made in Chapters 6 through 12. In particular, each of the two projection properties implies, by the results of Chapter 12 (Theorem 12.10 and Proposition 12.26) that K is a rectifiable set (see Chapter 11) or, equivalently, a regular set (see Chapter 8). The Uniform Density Property is connected with the notion of *Ahlfors set*.

Definition. *We say that a 1-set K is an* Ahlfors set *of Ω if there is a constant c such that, for every disk D contained in Ω and centered on K we have*

$$(15.30) \qquad\qquad c^{-1}R \leq \mathcal{H}^1(K \cap D) \leq cR.$$

The Uniform Density Property, which implies the left-hand inequality, combined with Property (M), which implies the right-hand inequality (proved in (14.6)), can be interpreted as an *a priori* bound on K among the Ahlfors sets. The same occurs with the Projection Properties, which are *a priori* bounds of *Uniform Rectifiability*. Thus, a natural choice of a suitable class of sets where to minimize the Mumford-Shah functional is at this point *the class $\mathcal{K}$ of the closed Ahlfors rectifiable subsets of Ω with a finite $\mathcal{H}^1$-measure.* One can easily observe that a minimum of the Mumford-Shah functional taken (if one exists!) in this class must enjoy the minimality property (M). An important point about the convenience of working in the class $\mathcal{K}$ is that all the elements therein can be approximated by finite unions of curves. In other words, this class will appear in the next theorem as the completion of a class of more regular objects. (Think of the analogy with the Sobolev space H^1, which is a natural completion of the space of all C^1-functions.) To be more precise, we shall now prove that a set in $\mathcal{K}$ can be covered by finitely many curves, with the exception of a small part.

Before this proof, a technical remark should be made. The estimates in this chapter are typical "interior estimates" and can only be applied to a disk $D(R)$ contained Ω. On the other hand, the existence and the approximation results which we are going to discuss are "global" and require the possibility of working uniformly on the whole domain. The problem can be easily overcome in at least three simple ways.

First, we can extend our estimates to boundary cases, as has been done, for instance, in [DalMMS2], with an elementary reflection technique which works when Ω is a rectangle. [DaSe3] treat this question for general domains. Alternatively, we can work on the subdomains of Ω, get an estimate on the length of the trace of the minimal set on all the compact subsets of Ω and finally pass to the limit. We shall adopt a third strategy, which avoids every further work and shows how irrelevant the boundary problem is: Since we are considering singular sets which do not intersect $\partial\Omega$, we shall add to all the set variables K the Ahlfors set $\partial\Omega$. This operation leaves the minimization problem invariant and makes the density estimates automatically satisfied near the boundary. Therefore, in this section, we shall always assume that K is a closed subset of $\overline{\Omega}$ with $\partial\Omega \subset K$.

Theorem 15.31. *Let $K \in \mathcal{K}$ and let $u = u_K$. Then, for every $\varepsilon > 0$, we can find a compact K_ε made of a finite number of rectifiable curves $c_1, c_2, ..., c_k$ and a function $u_\varepsilon \in C^1(\Omega \setminus K_\varepsilon)$ such that*

$$(15.32) \qquad\qquad \mathcal{H}^1(K \setminus K_\varepsilon) \le \varepsilon$$

$$(15.33) \qquad\qquad \sum_{j=1}^{k} l(c_j) \le \mathcal{H}^1(K) + \varepsilon$$

$$(15.34) \qquad\qquad I(u_\varepsilon, K_\varepsilon) \le I(u, K) + \varepsilon$$

$$(15.35) \qquad\qquad d(K, K_\varepsilon) \le \varepsilon.$$

(Note that, since

$$(15.36) \qquad\qquad \mathcal{H}^1(K_\varepsilon) \le \sum_{j=1}^{k} l(c_j),$$

from (15.33) and (15.34) we can deduce that

$$(15.37) \qquad\qquad E(u_\varepsilon, K_\varepsilon) \le E(u, K) + 3\varepsilon.)$$

Proof. By Theorem 11.23, since K is rectifiable, for all $\delta > 0$ we can find $H = \bigcup_{i=1}^{k} c_i$, where c_i are rectifiable curves, such that

$$(15.38) \qquad \sum_{j=1}^{k} l(c_j) \leq \mathcal{H}^1(K) + \delta \,,$$

$$(15.39) \qquad \mathcal{H}^1(K \setminus H) < \delta \,.$$

Since $\partial\Omega$ is a finite union of curves, we can also assume $\partial\Omega \subset H$. Let for all $x \in K \setminus H$, $r(x) = d(x, H)$. The balls $B(x, r(x))$ obtained in this way for x in the set $K \setminus H$ are contained in Ω and clearly form a covering of $K \setminus H$. We can apply Covering Lemma 7.1 in order to extract a subcovering $(B_i)_{i \in I}$, $B_i = B(x_i, r_i)$, such that the balls $B(x_i, \frac{r_i}{5})$ are disjoint. The Uniform Density Property implies (setting $c = \beta^{-1}$) that

$$\mathcal{H}^1(K \cap B(x_i, \frac{r_i}{5})) \geq \frac{1}{5c} r_i,$$

so, by (15.39),

$$(15.40) \qquad \begin{aligned} \sum r_i &\leq 5c \sum \mathcal{H}^1(K \cap B(x_i, \frac{r_i}{5})) \\ &= 5c\mathcal{H}^1(K \cap \bigcup_i B(x_i, \frac{r_i}{5})) \\ &\leq 5c\left(\mathcal{H}^1(K \setminus H)\right) \leq 5\delta c \,. \end{aligned}$$

Consider now $\tilde{H} = H \cup (\bigcup_i \partial B_i)$. We can redefine the curves c_j in order to make them cover all of the set $\tilde{H}$. In fact, for every value of the index i in the set I, ∂B_i meets H and therefore meets one of the curves c_j. Then we can modify the curve c_j by including in it the circle ∂B_i as a loop. One can iterate this operation recursively, including thus all the boundaries of the sets B_i. A passage to the limit which ensures the conclusion of this process is allowed by the Ascoli-Arzela Theorem, since at every recursion step we obviously have by induction, taking into account (15.40),

$$(15.41) \qquad \begin{aligned} \sum_{j=1}^{k} l(c_j) &\leq \mathcal{H}^1(K) + \delta + \sum_{i \in I} \mathcal{H}^1(\partial B_i) \\ &\leq \mathcal{H}^1(K) + (10\pi c + 1)\delta \,. \end{aligned}$$

Let $A = \Omega \cap (\bigcup_i B_i))$ and define a function v to be equal to zero on A and to u on $\Omega \setminus A$. It is clear that v is in $H^1(\Omega \setminus \tilde{H})$. Indeed, it is a restriction of u. Let us calculate the energy $I(v)$. Taking into account that $|g| \leq 1$,

$$
\begin{aligned}
I(v) &= \int_{\Omega \setminus \tilde{H}} |\nabla v|^2 + \int_{\Omega} (v - g)^2 \\
&= \int_{(\Omega \setminus K) \setminus A} |\nabla u|^2 + \int_{\Omega \setminus A} (u - g)^2 + \int_A g^2 \\
&\leq I(u) + \mathcal{L}_2(A) .
\end{aligned}
$$

(15.42)

Since the boundary of A has a length bounded by $10\pi c\delta$, as we see from (15.40), we can, given ε, fix a suitably small value of δ and then the set $\tilde{H}$ and the function v as above. Then we set $K_\varepsilon = \tilde{H}$ and $u_\varepsilon = v$ and we obtain (15.32) from (15.39), (15.33) from (15.41), and (15.34) from (15.42). Of course, we can impose that all curves c_j meet K, so that

$$
\mathbf{d}(K, K_\varepsilon) \leq \sup_j \mathcal{H}^1(c_j \setminus K) + \sup_i r_i .
$$

Thus the distance $\mathbf{d}(K, K_\varepsilon)$ can be estimated from (15.38) and (15.40) by $(5c + 1)\delta$. $\square$

The above theorem and, in particular, (15.37) show that we can take a minimizing sequence for the functional E on $\mathcal{K}$, namely a sequence $(K_n)_{n \in \mathbb{N}}$ such that

(15.43)
$$
\lim_n E(K_n) = \inf_{K \in \mathcal{K}} E(K) ,
$$

where each of the elements K_n is the union of a finite number n of curves. Moreover, (15.33) and (15.36) show that the sum $l(K_n)$ of the lengths of such curves can replace the term $\mathcal{H}^1(K_n)$ in the computation of $E(K_n)$ without affecting the validity of (15.43). Once this is done, we shall conclude that all uniform density properties discussed above hold for the K_n. We shall ask more, and replace in such a minimizing sequence the sets K_n by other sets satisfying the property (M). In order to realize this program we shall find convenient to regard the unions of finitely many curves simply as finite unions of closed connected sets.

Proposition 15.44. *Let $\mathcal{K}^m$ be the class of the sets which are unions of at most m closed connected sets. Then the minimum of the Mumford-Shah functional E on $\mathcal{K}^m$ is attained.*

Proof. Let K_n be a minimizing sequence for E in $\mathcal{K}^m$. Then, by Proposition 10.1, the sequence K_n admits a converging subsequence for the

Hausdorff distance, still denoted by K_n. In addition, Theorem 10.19 ensures that if $\tilde{K}_m$ denotes the limit of the sequence K_n,

$$\mathcal{H}^1(\tilde{K}_m) \leq \liminf_n \mathcal{H}^1(K_n),$$

and therefore, setting $\tilde{K}^m = \tilde{K}_m \setminus \partial\Omega$,

$$\mathcal{H}^1(\tilde{K}^m) \leq \liminf_n \mathcal{H}^1(K_n).$$

By Semicontinuity Lemma 13.6, we also have

$$I(\tilde{K}^m) \leq \liminf_n I(K_n),$$

and from the last inequalities we conclude that

$$\inf_{K \in \mathcal{K}^m} E(K) \leq E(\tilde{K}^m) \leq \liminf_n E(K_n) \to \inf_{K \in \mathcal{K}^m} E(K).$$

$\square$

In order to obtain a minimum for $E(K)$ in $\mathcal{K}$, we now let $m \to +\infty$ and wish to prove that $\tilde{K}^m$ tends to such a minimum.

Lemma 15.45. *The sets $\tilde{K}_m$ satisfy for various universal constants the Uniform Density Property, the Concentration Property and the First Projection Property. In addition, these properties hold on every disk $D(R)$ centered on $\tilde{K}^m$ and contained in Ω with $R \leq 1$.*

Proof. Since $E(\tilde{K}^m)$ is minimum in the class $\mathcal{K}^m$, and Since we took care in Definition 14.1 of excision to ensure that the excision by a disk contained in Ω should not increase the number of the connected components of K, we know that $\tilde{K}^m$ satisfies the property (M). Therefore, by Theorems 15.1, 15.5, 15.18 and 15.27, we know that $\tilde{K}^m$ satisfies the Essentiality Property (provided we remove a finite set of curves whose range is a single point), the Uniform Density, Concentration, and First Projection properties for various universal constants. $\square$

We now are in a position to prove one of the main results of this chapter.

Existence Theorem 15.46. *There exists a segmentation $\overline{K} \in \mathcal{K}$ such that $E(\overline{K}) = \inf_{K \in \mathcal{K}} E(K)$.*

Proof. Let $\overline{K}$ be one of the closed Hausdorff limits of $\tilde{K}_m$. Because the $\tilde{K}_m$ uniformly satisfy the Concentration Property, they satisfy (10.11) and we can apply Proposition 10.10 and Semicontinuity Theorem 10.14:

$$\mathcal{H}^1(\overline{K}) \leq \liminf_m \mathcal{H}^1(\tilde{K}_m) \leq \liminf_m l(\tilde{K}_m).$$

By Semicontinuity Lemma 13.6 we also have

$$I(\overline{K}) \le \liminf_m I(\tilde{K}_m) .$$

Combining the last two inequalities and using (15.43), we get

$$E(\overline{K}) = \inf_{K \in \mathcal{K}} E(K).$$

In order to end the proof of the thesis, we only have to show that $\overline{K} \in \mathcal{K}$. This follows from the fact that for every m, $\tilde{K}_m$ satisfies the First Projection Property, with a constant β_1 which does not depend on m. Then, Proposition 10.7 (applied to the projections of the sets $\tilde{K}_m$ on the sides of an arbitrary square) implies that $\overline{K}$ also enjoys the same property. This tells us, in particular, that the Uniform Density Property holds for $\overline{K}$ and therefore that $\overline{K}$ is an Ahlfors set. Then $\overline{K}$, satisfying Property (M), is rectifiable by Theorem 15.27. Thus $\overline{K}$ belongs to the class $\mathcal{K}$ and the proof is complete. $\quad\square$

6. Nonuniqueness of the Mumford-Shah minimizers.

We shall produce a generic example by showing that in many cases for g, and in some cases for the scale λ, the existence of two different minima can be proved. We consider an arbitrary image on $\mathbb{R}^2$, defined as a continuous function g with compact support. We introduce a scale parameter λ, like in Chapters 2 and 3, and we consider the functional

$$E_\lambda : K \to I(K) + \lambda \mathcal{H}^1(K).$$

For every λ we set $c_\lambda = \inf_{\mathcal{K}} E_\lambda$. The next lemma will permit an easy construction of two different minimal segmentations for the function g, one empty and the other one nonempty.

Lemma 15.47. *The function $\lambda \to c_\lambda$ is a continuous nondecreasing function such that*

$$(15.48) \qquad\qquad \lim_{\lambda \to 0} c_\lambda = 0$$

and $c_\lambda = c_\infty = I(\emptyset)$ for λ large.

Proof. The monotonicity of c_λ is obvious, and the continuity follows from the inequality

$$c_{\lambda'} \le E_{\lambda'}(K_\lambda) = E_\lambda(K_\lambda) + (\lambda' - \lambda)\mathcal{H}^1(K_\lambda) \le c_\lambda + |\lambda - \lambda'|\lambda^{-1}c_\lambda ,$$

where K_λ denotes a minimum of E_λ on $\mathcal{K}$. For the proof of (15.48), given $\varepsilon \geq 0$, we fix a 1-set K which divides the support of g in small squares on which the oscillation of g is smaller than ε. Then we have the bound

$$\forall \lambda : c_\lambda \leq E_\lambda(K) \leq \lambda \mathcal{H}^1(K) + \varepsilon \mathcal{L}_2(supp(g)),$$

from which (15.48) easily follows. The last part of the thesis follows from the fact that, for λ large, the minimum of E_λ is achieved on the empty set; namely we have $K_\lambda = \emptyset$. In fact, if x is in K_λ, by Uniform Density Property we see that

$$\mathcal{H}^1(K_\lambda) \geq \mathcal{H}^1(K_\lambda \cap B(x,1)) \geq \beta,$$

where β is the constant found in Theorem 15.5. Note that such a constant has been determined when the parameter λ is equal to 1 but that its value is not affected by having a larger λ. Therefore

$$\beta \leq \lambda^{-1} c_\lambda \leq \lambda^{-1} c_\infty = \lambda^{-1} I(\emptyset),$$

which yields an upper bound for λ. Thus $I(\emptyset) = c_\infty$. $\square$

Nonuniqueness Theorem 15.49. *Given any continuous image with compact support, there is at least a critical scale λ for which the Mumford-Shah functional has at least two minima.*

Proof. Let now λ^* be the least value of λ such that $c_\lambda = c_\infty$. Of course, we can take for K_{λ^*} the empty set since $E_{\lambda^*}(\emptyset) = I(\emptyset) = c_\infty = c_{\lambda^*}$. More generally, we can observe that, for $\lambda > \lambda^*$, $K_\lambda = \emptyset$. In fact, if we have $\lambda \geq \lambda^*$ with $\mathcal{H}^1(K_\lambda) \neq 0$ we have

$$c_{\lambda^*} \leq E_{\lambda^*}(K_\lambda) < E_\lambda(K_\lambda) = c_\lambda \leq c_\infty,$$

in contradiction with the definition of λ^*. On the other hand, when $\lambda < \lambda^*$, we must obviously have $K_\lambda \neq \emptyset$, since $E_\lambda(\emptyset) = c_\infty$. By Semicontinuity Theorems 10.14 and 13.6 again, we can find a minimum K^* of E_{λ^*}, taken as the Hausdorff limit of K_λ for some values of λ tending to λ^*, with $\lambda < \lambda^*$, and we would like to have $K^* \neq \emptyset$. If we take any square S with side length 1 centered at a point of K^*, by First Projection Property, when we pass to the limit, the sum of the projections of K_λ on the sides of S must be larger than β_1. By Proposition 10.7 the same property must be satisfied by K^*, so we obtain $\mathcal{H}^1(K^*) \neq 0$. Since $K = \emptyset$ also is a minimum of E_{λ^*}, we have found two different minima of the functional E_{λ^*}. $\square$

Chapter 16

FURTHER PROPERTIES OF MINIMIZERS: COVERING THE EDGE SET WITH A SINGLE CURVE

In this final chapter we shall consider further properties of the optimal segmentations. The main new information which we shall obtain is that the whole segmentation K is contained in a single rectifiable curve γ whose length is proportional (with a ratio depending on Ω) to the length of K and which is Ahlfors regular. More precisely, there is a universal constant C such that on every disk $D(r), r \leq 1$ one has $l(c \cap D(r)) \leq Cr$. The existence of such a curve will be obtained as a counterpart to a uniform rectifiability property, stronger than any considered in the last chapter. We shall prove that in every disk centered on K there is a rectifiable curve which is almost equal to a "big piece" of K. This property, which we call "Concentrated Rectifiability Property" will be proved in Sections 1 and 2 by using the "small oscillation covering" technique of Chapters 14 and 15. Section 3 is devoted to the proof that we can somehow join the curves given by this Rectifiability Property and obtain a curve containing all of K. In Section 4, we show by a simple minimality argument that such a curve can be imposed to be Ahlfors regular, with a universal constant only depending on the diameter of Ω.

1. The David-Semmes Concentrated Rectifiability Property.

In this chapter the notation γ will always indicate a rectifiable curve in the sense specified in Chapter 6 and in Section 3 of Chapter 11. We shall adopt the same notation for the curve and its range, so that we consider γ as a subset of the plane. However, this will lead to no ambiguity because $l(\gamma)$ denotes the length of the curve and $\mathcal{H}^1(\gamma)$ the Hausdorff measure of its range. In order to fix ideas and to avoid spurious geometric technicalities, we assume that Ω is a set whose boundary is a Jordan piecewise C^1 boundary.

Definition (Concentrated Rectifiability Property). *We shall say that K satisfies the Concentrated Rectifiability Property if, for every $\varepsilon > 0$, there exists $\alpha > 0$ such that for every $x \in K$ and for every $R \leq 1$ such that the disk $D(x, R)$ is contained in Ω, there is a curve γ satisfying*

$$\mathcal{H}^1(\gamma) \geq \alpha R \quad and \quad \mathcal{H}^1(\gamma \setminus K) \leq \varepsilon \mathcal{H}^1(\gamma).$$

As we did for the concentration and projection properties in the last chapter, we consider the property which characterizes sets not enjoying the Concentrated Rectifiability Property.

Definition (Atomized Curve Property). *Let x be a point of K and R such that the disk $D(x,R)$ is contained in Ω. We shall say that K satisfies the Atomized Curve Property on the disk $D(x,R)$, in correspondence with two given positive constants $\alpha > 0$ and $\varepsilon > 0$ if*

$$\forall \gamma \text{ such that } \mathcal{H}^1(\gamma) \geq \alpha R : \mathcal{H}^1(\gamma \setminus K) \geq \varepsilon \mathcal{H}^1(\gamma).$$

We shall prove in this section that the minimizers of the Mumford-Shah functional satisfy the concentrated rectifiability property. To this aim, we shall use a weaker version of the above property.

Definition 16.1 (Quantified Nonconnectedness Property). *Let x be a point of K and let $R \leq 1$ such that the disk $D(x,R)$ is contained in Ω. We shall say that K satisfies the Quantified Nonconnectedness Property on the disk $D(x,R)$, in correspondence of two given positive constants $\alpha > 0$ and ε if, for every $y \in D(x,R)$, for every positive constant $r \geq \alpha R$ such that $D(y,r) \subset D(x,R)$ and for every rectifiable curve γ connecting the two circles $\partial D(y, \frac{r}{2})$ and $\partial D(y,r)$, one has*

$$(16.2) \qquad\qquad \mathcal{H}^1(\gamma \setminus K) \geq \varepsilon r.$$

We point out that, if K satisfies the Atomized Curve Property, then the Quantified Nonconnectedness Property also holds with the same values for α and dividing ε by 2. The Quantified Nonconnectedness Property is illustrated in Figure 16.1.

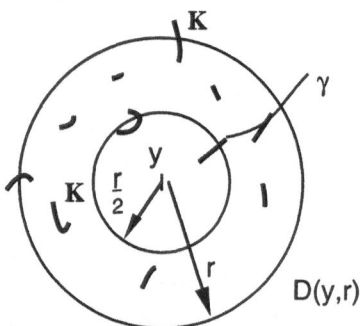

Figure 16.1 : Quantified nonconnectedness property

The next lemma holds for any closed set K and any nonnegative continuous function f defined on $D(y,r) \setminus K$, but the reader must be aware that the lemma will be applied with $f = |\nabla u|$ and a minimal segmentation K, so that (16.4) below will follow from (16.21).

Lemma 16.3. *Let K be a closed rectifiable set and assume that $y \in K$, r, $\varepsilon > 0$ are given in such a way that (16.2) is satisfied for every for every rectifiable curve γ connecting the circles $\partial D(y, \frac{r}{2})$ and $\partial D(y,r)$. Assume in addition that*

$$(16.4) \qquad \int_{D(y,r) \setminus K} f \leq C r^{\frac{3}{2}}.$$

Then there exists a connected and simply connected open set with a C^1-boundary V such that

$$(16.5) \qquad \overline{D}\left(y, \frac{r}{2}\right) \subset V \subset \overline{V} \subset \overline{D}(y,r)$$

$$(16.6) \qquad \partial V \quad \text{is connected}$$

$$(16.7) \qquad \partial V \cap K = \emptyset$$

$$(16.8) \qquad \mathcal{H}^1(\partial V) \leq \frac{C}{\varepsilon} r$$

$$(16.9) \qquad \int_{\partial V} f d\mathcal{H}^1 \leq \frac{C}{\varepsilon} r^{\frac{1}{2}},$$

where C is an absolute constant.

The proof of Lemma 16.3 will be the subject of the next section. We want to immediately show how the above result will be used in the line of the proofs developed in Chapter 15. In fact, Lemma 16.3 is a tool for constructing a small oscillation covering of K, whenever the Concentrated Rectifiability Property is not fulfilled.

Small Oscillation Covering Lemma. *Let K be a minimal segmentation, x a point of K and $R \le 1$ such that $D(x, R) \subset \Omega$. Assume that the Quantified Nonconnectedness Property is satisfied by K on $D(x, R)$ for two given positive constants ε and α. Then we can find a small oscillation covering, consisting of subsets with C^1-boundary $(V_i)_i$ of $K \cap D(x, R(1 - 2\alpha))$, with constants ν and μ proportional to ε^{-1} and $\sup_i p_i$ proportional to $\alpha \varepsilon^{-1}$.*

Proof. We can cover $K \cap D(x, R(1 - 2\alpha))$ with disks $D(y, r)$, centered on K and with radius $r = 2\alpha R$. Take $f = |\nabla u|$, so that for any disk $D(y, r)$, by Property (M) and the Cauchy-Schwarz Inequality, one has

$$\int_{D(y,r)} f \le \left(\int_{D(y,r)} |\nabla u|^2 \right)^{\frac{1}{2}} (\mathcal{L}_2(D(y,r)))^{\frac{1}{2}} \le C r^{\frac{3}{2}}.$$

By Lemma 16.3, we can construct for any $y \in K \cap D(x, R(1 - 2\alpha))$ a set V containing $D(y, \frac{r}{2})$ and having the properties (16.5)-(16.9). The sets V are a covering of $K \cap D(x, R(1 - 2\alpha))$ and we can, by using Covering Lemma 7.1 and the Uniform Density Property, proceed exactly as in the proofs of the concentration and projection properties. So we extract a small oscillation covering V_i of $K \cap D(x, R(1 - 2\alpha))$, with $\mu = \frac{C}{\varepsilon \beta}$ (β is the "elimination constant") and, by (16.9), $\nu = \frac{C}{\varepsilon}$. The announced bound on $\sup_i p_i$ is ensured by (16.8), since $r \le \alpha R$. $\square$

Theorem 16.10. *If K is a minimum of the Mumford-Shah functional, it satisfies the Concentrated Rectifiability Property on every disk $D(x, R)$ contained in Ω, centered on K and such that $R \le 1$. Moreover, the function which associates with ε the value of α is an universal function.*

Proof. In fact, if we can find a positive value of ε such that, whatever α is, the Quantified Nonconnectedness Property is satisfied on some disk $D(x, R)$, then we can construct a covering of $D(x, R(1 - 2\alpha))$ as in the above lemma and we get a contradiction with the statement of the Elimination Theorem (Theorem 14.15), provided α is small enough with respect to ε. $\square$

2. Dual curve lemmas.

We now return to the proof of Lemma 16.3, which follows the David-Semmes proof [DaS1]. The set V will be defined as a carefully chosen level set $\{x, \delta(x) \ge t\}$ of a Lipschitz function δ.

Lemma 16.11. *Let K and $D = D(y, r)$ be as in Lemma 16.3. Define $\delta(x)$ as the infimum of the values of $\mathcal{H}^1(\gamma \setminus K)$, extended to all the* *rectifiable curves* γ *contained in* $\overline{D}$ *and connecting* ∂D *to* x. *Then*

$$(16.12) \qquad \forall x, x' \in \overline{D} : |\delta(x) - \delta(x')| \leq |x - x'|,$$

$$(16.13) \qquad \delta(x) = 0 \quad if \quad x \in \partial D,$$

$$(16.14) \qquad \delta(x) \geq \varepsilon r \quad if \quad x \in \overline{D}(y, \frac{r}{2}).$$

Proof. In order to prove (16.12), let $x, x' \in \overline{D}$ and let γ be a rectifiable curve in $\overline{D}$ connecting x to ∂D. We then construct a curve γ' connecting x' to ∂D by simply adding to γ the segment from x to x'. Clearly, γ' is a curve in $\overline{D}$ and therefore $\mathcal{H}^1(\gamma' \setminus K) \leq \mathcal{H}^1(\gamma \setminus K) + |x - x'|$. By taking the infimum over all curves γ, we deduce that $\delta(x') - \delta(x) \leq |x - x'|$. The relation (16.13) is an obvious consequence of the definition of δ and the relation (16.14) follows from the Quantified Nonconnectedness Property of K. $\square$

Lemma 16.15. *Let* K *be as in Lemma 16.3. Then, for almost every* $t \geq 0$, *the "isolevel set"* $\delta^{-1}(t)$ *does not meet* K.

Proof. We want to prove that $\mathcal{H}^1(\delta(K)) = 0$. Since K is rectifiable, it can be covered (up to a negligible set Z) for every $\varepsilon > 0$ by a countable set of rectifiable curves γ_i such that $\sum_i \mathcal{H}^1(\gamma_i) \leq \mathcal{H}^1(K) + \varepsilon$. (Theorem 11.3.) Thus, by the definition of δ, we have for every x, y in γ_i, $|\delta(x) - \delta(y)| \leq \mathcal{H}^1(\gamma_i \setminus K)$ and therefore, for every index i, $\delta(\gamma_i)$ is an interval of $\mathbb{R}$ such that $\mathcal{H}^1(\delta(\gamma_i)) \leq \mathcal{H}^1(\gamma_i \setminus K)$. This implies $\mathcal{H}^1(\delta(K)) \leq \sum_i \mathcal{H}^1(\gamma_i \setminus K) \leq \varepsilon$ and therefore, since ε is arbitrarily small, $\mathcal{H}^1(\delta(K)) = 0$. $\square$

Lemma 16.15 has led us to a first important result: Whenever $K \cap D$ has the Atomized Curve Property, a lot of "dual curves", the sets $\delta^{-1}(t)$, can be found surrounding $D(y, \frac{r}{2})$ and not meeting K. Each one of these "curves" is a candidate to be the boundary of the wanted set V. Now, we also want properties (16.8) and (16.9), as a criterion of selection among the isolevel sets $\delta^{-1}(t)$. In particular, (16.8) means that the isolevel sets of δ cannot be all to long. If they were, δ should oscillate a lot, which would contradict its being Lipschitz. The formal translation of this intuition can be given thanks to the Coarea formula. We do not prove the formula in this book. A simple self-contained proof can be found in Evans-Gariepy ([EvG], Chapter 3.4, Lemma 2). See also the classical reference, Federer ([Fed], Theorems 16.2.11 and 16.2.12).

Coarea formula. *Let δ be Lipschitz function on $\mathbb{R}^2$ with Lipschitz constant less than 1 and, for every $t \geq 0$, denote by μ_t the restriction of $\mathcal{H}^1$ to $\delta^{-1}(t)$. Then we have the following inequality between measures*

$$\int_{\mathbb{R}} \mu_t \, dt \leq \mathcal{L}_2.$$

Proof of Lemma 16.3. Integrating the function f with respect to both measures yields

$$\int_0^{\varepsilon r} \int_{\delta^{-1}(t)} f \, d\mathcal{H}^1 \, dt \leq \int_D f \leq C r^{\frac{3}{2}}.$$

So we can use once more the argument already applied in the proof of Fubini Lemma 14.19 and in the previous chapter and we can find a suitable number $t \in (0, \varepsilon r)$ such that

(16.22) $$\int_{\delta^{-1}(t)} f \, d\mathcal{H}^1 \leq \frac{C}{\varepsilon} r^{\frac{1}{2}}.$$

In the same way we can also take t such that

(16.21) $$\mathcal{H}^1(\delta^{-1}(t)) \leq \frac{C}{\varepsilon} r, \quad \text{and}$$

and by Lemma 16.15 we can assume

(16.20) $$\delta^{-1}(t) \cap K = \emptyset.$$

Finally, (16.14) shows that, for such a value of t, we must have

(16.19) $$\overline{D}(y, \frac{r}{2}) \subseteq \{x \in \overline{D}, \quad \delta(x) \geq t\}.$$

We can now define V by taking the connected component U_0 of the set $U = \{x \in \overline{D}, \quad \delta(x) \geq t\}$ containing $D(y, \frac{r}{2})$ and by "filling its eventual holes". (In rigorous terms, V is the complementary set of the closure of the unbounded connected component of the complementary set of U_0.) Thus $\partial V \subseteq \partial U \subseteq \delta^{-1}(t)$ and so V satisfies all relations of Lemma 16.3. Finally the boundary of V can be made C^1 with little trouble. Indeed, since K is closed, we can slightly perturb the curve without touching K and, since f is continuous out of K, the integral estimates are preserved by suitably small perturbations. So we can make ∂V as smooth as we need. $\square$

3. Existence of a rectifiable curve containing the edge set.

We shall conclude this chapter with the most complete answer to the Mumford-Shah conjecture now available (January 1994).

Theorem 16.23. *Let Ω be a piecewise C^1 convex set. There exists a constant $C(\Omega)$ such that for every bounded "image" $0 \leq g(x) \leq 1$ defined on Ω and every closed and essential K minimizing the Mumford-Shah energy $E(K)$, one can find a curve γ containing all of K and satisfying $l(\gamma) \leq C\mathcal{H}^1(K \cup \partial\Omega)$ and*

$$l(\gamma \cap D(r)) \leq Cr$$

for any disk $D(r)$.

The essentiality assumption is not relevant, since we proved in the preceding chapter that from any minimal segmentation we can eliminate isolated points so that it becomes essential. The same occurs with closedness. We begin with a statement weaker than the Concentrated Rectifiability Property proved above. This statement will be enough for our aims.

Lemma 16.24. *There exist two universal constants c and C such that, for every disk $D(R)$ centered at a point of K and contained in Ω, one can find a rectifiable curve γ, contained in $\overline{D}(R)$, such that $\mathcal{H}^1(\gamma \cap K) \geq cR$ and $\mathcal{H}^1(\gamma) \leq CR$.*

Proof. This is a direct consequence of the Concentrated Rectifiability Property. We take $\varepsilon = \frac{1}{2}$, $\alpha = \alpha(\frac{1}{2})$, so that $\mathcal{H}^1(\gamma) \geq \alpha R$ and $\mathcal{H}^1(\gamma \setminus K) \leq \frac{1}{2}\mathcal{H}^1(\gamma)$. Thus

$$\mathcal{H}^1(\gamma \cap K) \geq \mathcal{H}^1(\gamma) - \mathcal{H}^1(\gamma \setminus K) \geq \frac{1}{2}\mathcal{H}^1(\gamma) \geq \frac{1}{2}\alpha R$$

and $\mathcal{H}^1(\gamma \setminus K) \leq \frac{1}{2}(\mathcal{H}^1(\gamma \setminus K) + \mathcal{H}^1(\gamma \cap K))$, so that

$$\mathcal{H}^1(\gamma \setminus K) \leq \mathcal{H}^1(\gamma \cap K).$$

Therefore

$$\mathcal{H}^1(\gamma) \leq 2\mathcal{H}^1(\gamma \cap K) \leq 2\mathcal{H}^1(K \cap D(R)) \leq 6\pi R.$$

$\square$

Remark. Of course, we can assume that the curve γ found above is a loop and that $\partial D(R) \subset \gamma$. In this case, the value of the constant C must be modified by adding $2\pi+1$. Moreover, if we start by using Lemma 16.24 in a sligthly smaller disk, we can also assume that the estimate $\mathcal{H}^1(\gamma \cap K) \geq cR$ becomes $\mathcal{H}^1(\gamma \cap K \cap D(R)) \geq cR$.

Proposition 16.25. *There exists a universal constant C' such that, for every disk $D(R)$, $R \leq 1$ centered at a point of K, one can find a rectifiable curve γ, contained in $\overline{D}(R)$, such that $K \cap \overline{D}(R) \subset \gamma$ and*

$$(16.26) \qquad\qquad H^1(\gamma) \leq C' \mathcal{H}^1(K \cap \gamma).$$

Proof. Fix $C' = \frac{C}{c}$, where the constants c and C are the same as in Lemma 16.24, modified according to the above remark. Consider the class Γ of the curves γ contained in $\overline{D}(R)$, satisfying (16.26) and containing $\partial D(R)$. We now claim that a maximal element by inclusion in Γ must contain all of $K \cap \overline{D}(R)$. For, if γ is maximal and if, by contradiction, there is some point $x \in (K \cap D(R)) \setminus \gamma$, we set $r = \mathbf{d}(x, \gamma)$ and we take, by the above remark, a loop γ_r in $\overline{D}(x, r)$ such that

$$\partial D(x, r) \subset \gamma_r, \quad \mathcal{H}^1(\gamma_r) \leq \frac{C}{c} \mathcal{H}^1(\gamma_r \cap K \cap D(x, r)).$$

Since γ_r is a loop which meets γ, we can insert it in the curve and obviously $\gamma \cup \gamma_r \in \Gamma$, in contradiction with the maximality of γ.

Finally, we must show how to find a maximal element in Γ. We know by Lemma 16.24 that Γ is not empty. This allows us to define an increasing sequence by inclusion $(\gamma_n)_{n \in \mathbb{N}}$, starting from any element of Γ and proceeding by recursion in the following way. After we have chosen the n-th element γ_n of the sequence, we denote by s_n the supremum of the $\mathcal{H}^1$-measures of all the curves in Γ which strictly contain γ_n (provided γ_n is not already a maximal element) and we choose as γ_{n+1} a curve in Γ containing γ_n and with a measure larger than $s_n - \frac{1}{n}$. Since the sequence consists of curves with bounded lengths because of (16.26), we can pass to the limit by using the Ascoli-Arzela Theorem. By monotonicity we find a curve γ which contains all the curves γ_n.

We can easily prove the maximality of γ. In fact, if γ' is another curve in Γ which strictly contains γ, then it has a stricly larger measure and therefore a measure strictly larger than s_n, for n large. This contradicts the definition of s_n because γ' is in Γ and contains γ_n. $\square$

4. Existence of a regular curve containing the edge set.

For technical reasons, we shall find more convenient to regard the curve γ found in Proposition 16.25 as an element of the more general class of the connected sets contained in $\overline{D}(R)$.

Lemma 16.27. *Under the assumptions of Proposition 16.25, there is a continuum T contained in $\overline{D}(R)$, containing $K \cap \overline{D}(R)$, and regular. Namely, we can impose that for any subdisk $D(y,r)$ contained in $D(R)$,*

$$(16.28) \qquad \mathcal{H}^1(T \cap D(y,r)) \leq 4\pi C'r,$$

where C' is the constant of (16.26).

Proof. By Semicontinuity Theorem 10.19, we can find an element T with minimal $\mathcal{H}^1$-measure in the class of the closed connected subsets of $\overline{D}(R)$ such that $K \cap \overline{D}(R) \subset T$ and

$$\mathcal{H}^1(T) \leq C'\mathcal{H}^1(K \cap T) = \mathcal{H}^1(K).$$

Of course, the preceding class has at least one element by Proposition 16.25. We are going to show that the minimal set satisfies (16.28).

In fact, if x and r are such that (16.27) does not hold, we can take in $\overline{D}(x,r)$ a curve γ_r as in Proposition 16.25, that is, containing all of $K \cap \overline{D}(R)$ and satisfying

$$\mathcal{H}^1(\gamma) \leq C'\mathcal{H}^1(K \cap \gamma) \leq 3\pi C'r.$$

The last inequality is true because K satisfies the property (M). Then we make an excision, passing from T to the set $T' = (T \setminus D(x,r)) \cup \gamma \cup S$, where S is some segment inside $D(x,r)$ connecting γ to T. Thus, such a set still is connected and

$$\mathcal{H}^1(T') \leq \mathcal{H}^1(T) - 4\pi C'r + 3\pi C'r + 2r < \mathcal{H}^1(T),$$

because $2 < \pi C'$, which yields a contradiction. $\qquad \square$

The above argument shows the reason for which we have preferred to show the regularity condition for a minimal connected set containing K rather than for a minimal curve. As we have seen in Chapter 14, the excision operation preserves the property of being a connected set but not that of being a curve. However, there is not much difference between a closed connected set with finite length and a rectifiable curve. We shall draw from the next well-known result that by simply multiplying the constants by 2, the result of the preceding lemma still holds for some rectifiable curve.

Lemma 16.29. *Every closed connected set T with finite $\mathcal{H}^1$-measure is equal to the range of a rectifiable curve γ such that almost every point in T is assumed by γ at most twice. As a consequence,*

$$(16.30) \qquad\qquad l(\gamma \cap D(r)) \leq 2\mathcal{H}^1(T \cap D(r))$$

for any disk $D(r)$.

Proof. We take a curve γ passing at most twice by almost every point of T, satisfying (16.30) and maximal by inclusion. We wish to prove that γ contains all of T. Assume by contradiction that we can find a point $x \in T$ out of γ. Then there is a path γ_1 contained in $T \setminus \gamma$ and joining x to γ. Indeed, by Theorem 6.33, T is path connected. Thus we can include $\gamma_1 \cup \tilde{\gamma}_1$ as a loop of γ and (16.30) is clearly preserved. In fact, we have added double points to γ with the exception of the contact point of γ_1 and γ. Thus we get a contradiction. That some maximal curve γ can be found results from the same argument as in Theorem 16.24. □

Proof of Theorem 16.23. We can notice that everything we did with the assumption $R \leq 1$ works all the same (with constants depending on $\operatorname{diam}(\Omega)$ for $R \leq \operatorname{diam}(\Omega)$). By using as usual Covering Lemma 7.1 and the Uniform Density Property (15.4), we can cover K with disks $D_i(r_i)$ centered on K and tangent to $\partial\Omega$, satisfying $\sum_i r_i \leq 5\beta^{-1}\mathcal{H}^1(K)$. By Lemma 16.25, and the remark preceding it, we see that for every i we can cover $K \cap D_i$ with a loop γ_i touching $\partial\Omega$ and satisfying $l(\gamma_i \leq Cr_i)$. All γ_i can be inserted in the Jordan curve $\partial\Omega$ as loops, so that we have constructed a curve containing $K \cup \partial\Omega$ with length less than $C\mathcal{H}^1(K) + l(\partial\Omega)$. In order to construct such a curve which in addition is Ahlfors regular, we can use without change the argument of Lemma 16.27. □

Bibliographical notes

As we have seen, image segmentation can be viewed under many aspects and we can only apologize for the necessary uncompleteness of the references. They are classified as follows.

References of Part I.
I-A) Image segmentation and edge detection, surveys and monographs.
I-B) Articles proposing algorithms for edge detection and image segmentation.
I-C) Scale space theory.
I-D) Mathematical analysis related to scale space theory.
I-E) Monographs in Image Processing.
I-F) Articles on texture analysis and segmentation.
I-G) Wavelets, theory and relation to image processing and scale space.
I-H) Related topics in psychophysics, neurobiology and Gestalt theory.

References of Part II.
II-A) References on geometric measure theory and rectifiability.
II-B) Monographs on mathematics and geometric measure theory.

References of Part III.

Notes on Part I.

In Part I of this book, we have focused on the image processing and algorithmic viewpoint. Most of the references quoted in Part I are contained in I-A to I-E. All papers discussing grey-level segmentation are listed in I-B. Image segmentation surveys are reassembled in I-A. These ten surveys were all very useful to establish the classification of image segmentation algorithms made in Part I, particularly [Zu1, 2]. We have thought it useful to add in I-F a series of papers on texture analysis which we have consulted for completeness (because the general segmentation problem consists of segmenting an image in regions which have homogeneous textures rather than homogeneous grey-levels). In the same way, we have separated, but maintained in I-G, a small list of references on wavelets, keeping the part of it which proves close in spirit to the edge detection problem. Indeed, edge detection can be done by multiscale linear filtering, as was explained in 2.1, and fast linear multiscale filtering has been recently identified with wavelet theory. This relation is explained in [Mall], [MallZ], [Mey], [Fro], [Hum]. In [DVL], one finds an interesting comparison between some wavelet decomposition models for images and the Mumford-Shah model. [IkU] (and many references in the bibliography on textures, particularly [BecSI], [CogJ], [JuK], [MaliP], [VoP2]) discuss what the "best wavelet" is for texture analysis, a texture being understood as a particular statistical distribution of tiny edges.

I-B gives a list of references dealing with edge-detection and segmentation algorithms. This is just a small selection, more than a thousand algorithms for edge-detection and segmentation having been proposed in the past thirty years. The references in I-C and I-D are much more developed than what has been exposed on scale space in Part I. We have only explained in this book the part of scale space theory which could be useful for the understanding of the edge-detection and segmentation theory. I-H gives a list of references in psychophysics and neurobiology which we consider relevant for the edge-detection problem. Some of them, like [Kani], have been quite influencial for the design of segmentation algorithms. In the same way, the famous Hubel and Wiesel neurobiological paper [HubW] has had a strong effect on the edge-detection ideology.

Part II.

While we thought it useful and necessary to discuss and justify the relevance of variational methods in image segmentation, we by no means think that the mathematical analysis of the Mumford-Shah functional needs this justification in order to be performed. As we explained in the introduction, the basic question leading to the Mumford-Shah functional is the problem of characterizing planar sets of points which behave like a bunch of lines (the rectifiable sets). The complete general mathematical analysis of such sets was performed by Besicovitch in a series of three papers ([Besi1], [Besi2], [Besi3]), using the Carathéodory and Hausdorff definitions of length and dimension ([Cara], [Haus]). The Besicovitch theory is fully explained in the now classical Falconer book [Fal1]. Besicovitch proved the equivalence between three properties which have been considered in Part II of this book: regularity, rectifiability and projection property. Federer [Fed1] generalized the equivalence between rectifiability and projection property in arbitrary dimensions, and Marstrand [Mars1] and Mattila [Matt1] proved the equivalence between regularity and rectifiability in any dimension. Preiss [Preis] did a further generalization by proving that the mere existence almost everywhere of a density implies the rectifiability. In Part II, we developed the theory following the techniques due to Besicovitch, Marstrand and Mattila, because they are extremely suggestful for image processing. The Marstrand and Mattila proofs, particularly the reflection technique, gives a new light even in the case of 1-sets of the plane. Complete references, and a complete account of the state of the art in the geometry of Hausdorff measurable sets are given in Mattila's recent monograph [Matt2].

In Chapter 10, we explained a generalization of Golab's compactness theorem which proved very useful in the analysis of the Mumford-Shah functional. We took this generalization from [DalMMS2].

The Mumford-Shah conjecture.

This conjecture was first stated in [MumS1] (1985), in a conference paper which was later arranged and published [MumS2] (1988). In [MumS3], Mumford and Shah performed a very complete argumentation in favour of their conjecture, including the proof of the piecewise constant case, statement of the Euler equations, proof of the conjecture under *a priori* regularity assumptions and study of the relation with fracture theory. (This last relation was also discussed in [BlatM2] using ideas coming from crack propagation theory [Kn].)

Two simple constructive proofs of the Mumford-Shah conjecture in the case of piecewise constant approximation functions were given in [Wan] and [MorS1]. The last mentioned proof led to a fast segmentation algorithm which was proposed in [KLM] and developed for practical applications by the firm Cognictec Inc. ([RuKo], [ROF]). Lenny Rudin, at Cognitec Inc., used the segmentation algorithm for his expertise in the Los Angeles riots trials ([RuOs] 1993). The results of the algorithm were admitted by the Court as evidences.

An affine invariant version of the Mumford-Shah functional was proposed in [BaCG] [BaGo]. The Mumford-Shah functional in the case of piecewise constant approximates has also been studied in higher dimension in [CoTa1], [CoTa], [MaTa], [MorS4], where existence, density results as well as estimates on the number of regions are obtained.

The first idea for solving a weakened version of the general Mumford-Shah conjecture came from De Giorgi [DeGiAm], [DeGi0], [DeGi1], who proposed in 1988 to model the solution (u, K) of the problem as a "Special Bounded Variation" function. The formalism was not only aimed at modelling image segmentation, but also and mainly problems arising in Continuum Mechanics from the study of drops of liquid crystals. Function with bounded variation, or "BV functions" are functions whose derivative is a Radon measure. It has been proved that their discontinuity set is a rectifiable set (see [Giu], [Zie], II-B), so that a rectifiable set may conversely be described as the set of jumps of some BV function. De Giorgi proposed to use this property to describe the discontinuity set K in the Mumford-Shah functional as the edge set of the approximating function u. In order to modelize the smoothness of u outside K, one imposes the derivative of u outside K to be an integrable function. De Giorgi and Ambrosio called such functions "Special functions with Bounded Variation" (SBV functions). Subsequently, Ambrosio gave the first existence proof for SBV minimizers of the Mumford-Shah functional [Amb1], [Amb2]. An excellent review of the SBV-approach with application to the Mumford-Shah conjecture is [Amb3].

It is worth noticing that an anterior paper to propose BV functions as an adequate model for images is [Ru] (1987). Using ideas coming from Vol'pert [Vol] Rudin noticed that if the space of functions with bounded variation (BV) was adopted as the adequate space for images, the set of jumps of the function could be identified with the edge set.

Let us return to the progress in the proof of the Mumford-Shah conjecture. The Ambrosio results implied that K was rectifiable, but no more quantitative or qualitative information on K was available until De Giorgi, Carriero and Leaci [DeGiCL] proved that K is a closed set (and therefore could not for example be dense!). Several successive surveys on the "free discontinuity problems" have been written by De Giorgi [DeGi0, 1, 2, 3].

Not much progress has been done since in the case of images with arbitrary dimensions. In this book, we focused on the two-dimensional case, on which much more information is now avalaible. In Chapters 13 through 16, we followed the approach developped in [DalMMS1, 2], which furnishes a unifying view on further developements of the subject. The line of these chapters is to obtain progressively sharper estimates about the density and rectifiability of the edge sets minimizing the Mumford-Shah functional. The Uniform Density and Concentration properties were first proved in [DalMMS2]. The Uniform Projection Property was announced in [DibK] and proved in [Dib]. An alternative proof was published by Léger [Lég].

The Projection Property can be understood as a quantified rectifiability property. It is, however, not equivalent to the definitions of uniform rectifiability given in [DaSe1] by David and Semmes. In the same way, Jones [Jon] noticed that the continuous solution to the travelling salesman problem assumed the set of points by which the salesman has to pass to be rectifiable and proposed new rectifiability criteria. Chapter 16 gives an account of David and Semmes [DaSe3] results on the Concentrated Rectifiability of the edge set in the Mumford-Shah model. They do prove more than what is stated in Chapter 16. Now, to state their exact results would have led to introduce more formalism, namely about Carleson measures, and we thought it best to prove the main David and Semmes result by the techniques already established for the density properties. This main result states that the edge set K is contained in a single regular curve.

While this book was being the edited (summer '94), we learned of several new results about the Mumford-Shah conjecture. Guy David (personal communication) proved that there is a constant c such that every disk $D(x, r)$ centered on the segmentation K contains a piece of K with length larger than cr, which is a C^1 curve. This result is a (stronger) analogue of the "concentrated rectifiability property" proved in Chapter 16. Alexis Bonnet (personal communication) proved that if we impose that K has a finite number of connected components, then the minimizing K is made of a finite number of C^1 curves. Both results imply that K is smooth (C^1 and more) except for a set with zero Hausdorff measure. The results and techniques developed in this book are useful (and, in a large extent, necessary) for understanding the David and Bonnet work.

Related mathematical topics.

The mathematical analysis of the Mumford-Shah functional has led mathematicians to propose proved algorithms for minimizing the functional. Among such attempts, one can mention [AmbTo1], [AmbTo] and [BPV]. The two first mentioned works relate the minimization method for the Mumford-Shah functional to nonlinear diffusion systems. This way of thought was further developed in [Rich1], [Sha1], [Sha2]. In [Cham], the Γ-convergence of many classical discrete segmentation algorithms towards the Mumford-Shah functional was rigorously proved. The proof uses the recently proved fact that minimizers of the functional can be approximated by sets of piecewise linear edges [DiSé]. All the mathematical work developed on drops of liquid crystals and phase transition is very close in spirit to what has been discussed in this book. Let us mention [Mod] as a good introduction to phase transition problems. In a series of papers ([CaLe], [CaLe1], [CaLT, 1, 2], [Le0, 1, 2]), Carriero, Leaci and Tomarelli have developed the free discontinuity paradigm with applications to Elasticity and Plasticity problems. They study generalizations of the Mumford-Shah functional (also considered in segmentation: see Blake and Zisserman [BlakZ] in I-A), which apply to elastic plates subject to bending as well as to fracturing.

References

References of Part I.

I-A) Image segmentation and edge detection, surveys and monographs.

[Bi2]   T.O. Binford, *Survey of model-based image systems*. International Journal of Robotics Research, 1(1), 18-64, Spring 1982.

[BlakZ]   A. Blake and A. Zisserman, *Visual Reconstruction*.MIT Press, 1987.

[Brad1]   M. Brady, *Computational approaches to Image Understanding*. ACM Comput. Surveys, 14(1), 3-71, March 1982.

[Brad2]   M. Brady, *Criteria for representation of shape*. In: *Human and Machine Vision*, Beck and al. eds, Academic Press, New York-Orlando, FL, 1983.

[Dav]   L. Davis, *A survey of edge detection techniques*. Computer Graphics and Image Processing, 4, 248-270, 1975.

[FuM]   K.S. Fu and J.K. Mui, *A survey on image segmentation*. Pattern Recognition 13, 3-16, 1981.

[HarS]   R.M. Haralick and L.G. Shapiro, *Image segmentation techniques*. Computer Vision Graphics and Image Processing, 29, 100-132, 1985.

[Pav3]   T. Pavlidis, *A critical survey of image analysis methods*. IEEE Proc. of the 8th Int. Conf. on Pattern Recognition, Paris, 1986

[Wesz]   J.S. Weszka, *A survey of threshold selection techniques*. Computer Graphics and Image processing, 7, 259-265, 1978.

[Zu1]   S.W. Zucker, *Region growing: Childhood and Adolescence (Survey)*. Comp. Graphics and Image Proc. 5, 382-399, 1976.

[Zu2]   S.W. Zucker, *Algorithms for image segmentation*. In: "Digital Image Processing and Analysis", ed. J.C. Simon and A. Rosenfeld, 1977.

I-B) Articles proposing algorithms for edge detection and image segmentation.

[ALM]   L. Alvarez, P.-L. Lions and J.M. Morel, *Image selective smoothing and edge detection by nonlinear diffusion (II)*, SIAM Journal on numerical analysis **29**, 845–866, 1992.

[AmiTW]   A. Amini, S. Tehrani and T. E. Weymouth, *Using dynamic programming for minimizing the energy of active contours in the presence of hard constraints*. Second International Conference on Computer Vision (Proceedings of), ICCV'88, IEEE n° 883.

[Az1]   R. Azencott, *Synchronous Boltzmann machines and Gibbs fields; Learning algorithms*. To appear, Springer Verlag 1989, (NATO Series, Les Arcs Congress Proceedings).

[Az2] R. Azencott, *Boltzmann Machines: high order interactions and synchronous learning.* Submitted to IEEE PAMI Trans.

[BajLL] R. Bajcsy, S.W. Lee and A. Leonardis, *Color image segmentation with detection of highlights and local illumination induced by inter-reflections.* IEEE Proc. 10th Int. Conf. on Pattern Recognition (I), 785-790, Atlantic City 1990.

[BajR] R. Bajcsy and D.A. Rosenthal, *Visual and conceptual focus of attention.* In: Structured Computer Vision, 133-149, S. Tanimoto and A. Klinger, eds., New York, Academic, 1980.

[BajSG] R. Bajcsy, F. Solina and A. Gupta, *Segmentation versus object representation - are they separable?* In: *Analysis and interpretation of range images.* R. Jain and A.K. Jain eds, Springer Verlag, New York, 1990.

[BajT] R. Bajcsy and M. Tavakoli, *Computer recognition of roads from satellite pictures.* IEEE Trans. Syst., Man, Cybern., SMC-6(9), 623-637, 1976.

[Baj] R. Bajcsy, *Computer identification of visual surfaces.* Computer Graphics and Image Processing, 2, 118-130, 1973.

[BeaG] J.M. Beaulieu and M. Goldberg, *Hierarchy in picture segmentation: a stepwise optimization approach.* IEEE PAMI, 11(2), February 1989.

[BergerM] M.O. Berger and R. Mohr, *Towards autonomy in active contour models.* IEEE Proc. 10th Int. Conf. on Pattern Recognition (I), 847-851, Atlantic City 1990.

[Berz] V. Berzins, *Accuracy of Laplacian edge detectors.* Computer Graphics and Image Processing, 27, 195-210, 1984.

[BeslJ] P.J. Besl and R. Jain, *Segmentation through variable-order surface fitting.* IEEE PAMI, 10, n°2, 167-192, March 1988.

[Besl] P.J. Besl, *Surfaces in range image understanding.* Springer Verlag, 1988.

[BevGKHR] J.R. Beveridge, J. Griffith, R.R. Kohler, A.R. Hansen and E.M. Riseman, *Segmenting images using localized histograms and region merging.* Tech. Report 87-88, Dept. of Computer and Information Science, Univ. of Massachusetts, Amherst, 1987.

[Bh] B. Bhanu, *Representation and shape matching of 3-D objects.* IEEE PAMI 6(3) 340-350, May 1984.

[Bi1] T.O. Binford, *Inferring surfaces from images.* Artificial Intelligence, 17, 205-244, 1981.

[BolF] R.C. Bolles and M. A. Fischler, *A paradigm for model fitting with applications to image analysis and automated cartography.* Comm. ACM, 24(6), 381-395, June 1981.

[BorP] A. Borsellino and T. Poggio, *Correlation and convolution algebras.* Kybernetic, 13, 113-122, 1973.

[BracS] R. Bracho and A.C. Sanderson, *Segmentation of images based on in-*

tensity gradient information. Proc. CVPR-85, 341-347, 1985.

[BriF] C. Brice and C. Fennema, *Scene analysis using regions.* Artificial Intelligence, 1, 205- 226, 1970.

[BrLe] L. Brenac and J. Lemaire, *Segmentation hiérarchique d'images en maillage hexagonal.* Technical Report 93-32, LISSS, Sophia-Antipolis, 1993.

[Bro] M.J. Brooks, *Rationalizing edge detectors.* Computer Graphics and Image Processing, 8, 277-285, 1978.

[Can] J. Canny, *A computational approach to edge detection.* IEEE Trans. PAMI 8(6) 679- 698, 1986.

[CatCLM] F. Catté, T. Coll, P.-L. Lions et J.M. Morel, *Image selective smoothing and edge detection by nonlinear diffusion,* SIAM Journal on numerical analysis, 1991.

[CCCD] V. Caselles, F. Catté, T. Coll and F. Dibos, *A geometric model for active contours in image processing* Report 9210, CEREMADE. Université Paris Dauphine, 1992.

[Ch] D.S. Chen, *A data-driven intermediate level feature extraction algorithm.* IEEE PAMI, 11(7), 749-758, July 1989.

[CoGe] G.H. Cottet and L. Germain, *Image processing through reaction combined with nonlinear diffusion.* Mathematics of Computation, 61, 204, 659-673, 1993.

[CohVSG] L. Cohen, L. Vinet, P. T. Sander and A. Gagalowicz, *Hierarchical region based stereo matching.* Preprint, INRIA-Rocquencourt.

[CooEl] D.B. Cooper and H. Elliot, *A maximum likelihood framework for boundary estimation in noisy images,* in IEEE Trans. Pattern Recognition Machine Intelligence, PAMI-1, 372-384, 1979.

[Cu] S.R. Curtis, *Reconstruction of multidimensional signals from zero-crossings.* Res. Lab. Electron., MIT, Cambridge, Tech. Rep. 509, 1985.

[De] R. Deriche, *Using Canny's Criteria to derive a recursively implemented optimal edge detector.* Int. J. of Comp. Vision, 167-187, 1987.

[Fair] J. Fairfield, *Toboggan contrast enhancement for contrast segmentation.* IEEE 10th Conf. on Pattern Recognition, 712-716, 1990.

[FauH] O.D. Faugeras and M. Hebert, *The representation, recognition and location of 3-D objects.* Int. J. Robotics Research, 5(3), 27-52, Fall 1986.

[FelY] J.A. Feldman and Y. Yakimovsky, *Decision theory and artificial intelligence: 1. A semantics based region analyzer.* Artificial Intelligence, 5, 349-371, 1974.

[Fre] E.C. Freuder, *Affinity: a relative approach to region finding.* Computer Graphics and Image Processing, 5, 254-264, 1976.

[GagM] A. Gagalowicz and O. Monga, *A new approach for image segmentation.* Proceedings of the Eigth International Conference on Pattern Recognition, Paris, 265-267, IEEE, 1986.

[GeiY] J.D. Geiger and A. Yuille, *A common framework for image segmenta-*

tion. IEEE Proc. 10th Int. Conf. on Pattern Recognition (I), 502-507, Atlantic City 1990.

[GemG]   S. Geman and D. Geman, *Stochastic relaxation, Gibbs distributions and the Bayesian restoration of images.* IEEE PAMI 6, 1984.

[GoS]   M. Goldberg and S. Shlien, *A four dimensional histogram approach to the clustering of Landsat-data.* Mach. Process. of Remotely Sensed Data. IEEE CH 1218-7 MPRSD, 250-259, 1977.

[GriP]   W.E.L. Grimson and T. Pavlidis, *Discontinuity detection for visual surface reconstruction.* Computer Vision, Graphics, Image Processing, 30, 316-330, 1985.

[Gri]   W.E.L. Grimson, *From images to surfaces.* Cambridge, MA, MIT Press 1981.

[Guz]   A. Guzman, *Decomposition of a visual scene into three dimensional bodies.* Proc. Fall Joint Comp. Conf., 33, 291-304, 1968.

[Har2]   R.M. Haralick, *Digital step edges from zero crossing of second directional derivative.* IEEE PAMI, 6, 58-68, 1984.

[HoroP1]   S.L. Horowitz and T. Pavlidis, *Picture segmentation by a directed split-and-merge procedure.* Proc. Second Int. Joint Conf. Pattern Recognition, 424-433, 1974.

[HoroP2]   S.L. Horowitz and T. Pavlidis, *Picture Segmentation by a Tree Traversal Algorithm.* Journal of the ACM 23, 368-388, 1976.

[Hue]   M.H. Hueckel, *A local visual operator which recognizes edges and lines.* J.A.C.M., 20, 634-647, October 1973.

[JarP]   R.A. Jarvis and E.A. Patrick, *Clustering using a similarity measure based on shared near neighbors.* IEEE Trans. Comput. C-22, 1025-1034, 1973.

[Kana]   T. Kanade, *Region segmentation: signal vs. semantics.* Computer Graphics and Image Processing, 13, 279-297, 1980.

[KasWT]   M. Kass, A. Witkin and D. Terzopoulos, *Snakes: active contour models.* 1st Int.Comp.Vis.Conf., IEEE n°777, 1987.

[KatLP]   N. Katzir, M. Lindenbaum and M. Porat, *Planar curve segmentation for recognition of partially occluded shapes.* IEEE Proc. 10th Int. Conf. on Pattern Recognition (I), 842-846, Atlantic City 1990.

[KerMWP]   D. Kern, R. Marcus, M. Werman and S. Peleg, *Segmentation by minimum length encoding.* IEEE Proc. 10th Int. Conf. on Pattern Recognition (I), 681-683, Atlantic City 1990.

[KetL]   R.L. Kettig and D.A. Landgrebe, *Classification of multispectral image data by extraction and classification of homogeneous objects.* IEEE Trans. Geosci. Electron., 14(1), 19-26, 1976.

[KLM]   G. Koepfler, C. Lopez, J.M. Morel, *A multiscale algorithm for image segmentation by variational method,* to appear in SIAM journal on Numerical Analysis, 1993.

[KoeMS] G. Koepfler, J.M. Morel and S. Solimini, *Segmentation by minimizing a functional and the "merging" methods.* Cahiers de Mathématiques de la Décision, n° 9022,1990; presented in september 1991 at the '*GRETSI* Colloque' in Juan-les-Pins (France).

[KWT] M. Kass, A. Witkin and D. Terzopoulos, *Snakes: active contour models,* 1st Int.Comp.Vis.Conf. IEEE 777, 1987 .

[LeBSFB] E. Le Bras-Mehlman, M. Schmitt, O.D. Faugeras et J.D. Boissonat, *How Delaunay triangulation can be used for representing stereo data.* ICCV 1988.

[Lec] Y. Leclerc, *Constructing stable descriptions for image partitioning.* Int. J. C. Vis. 3, 73- 102, 1989.

[LeeP] D. Lee and T. Pavlidis, *One-dimensional regularization with discontinuities.* IEEE PAMI, 10, 822-829, November 1988.

[LeoGB] A. Leonardis, A. Gupta and R. Bajcsy, *Segmentation as the search for the best description of images in terms of primitives.* Technical report n° MS-CIS-90- 30, GRASP LAB 215, University of Pennsylvania.

[LeviL] M.D. Levine and J. Leemet, *A method for non-purposive picture segmentation.* Proc. of the Third International Joint Conference on Pattern Recognition, 1976.

[Mali] J. Malik, *Interpreting line drawings of curved objects.* International Journal of Computer Vision, 1(1), 73-103, 1987.

[MarrHi] D. Marr et E. Hildreth, *Theory of edge detection.* Proc. Roy. Soc. Lond. B207, 187-217, 1980.

[Marro] J.L. Marroquin, *Probabilistic solution of inverse problems.* PhD thesis, MIT Cambridge, MA, September 1985.

[MarrP] D. Marr and T. Poggio, *A computational theory of human stereo vision.* Proc. Roy. Soc. London, B 207, 301-328, 1979.

[Mart] A. Martelli, *Edge detection using heuristic search methods.* Computer Graphics Image Processing 1, 169-182, 1972.

[MaRu] N. Mankovitch, L. Rudin et al., *Applications of a new pyramidal segmentation algorithm to medical images.* Proc. SPIE: Image processing Medical Imaging VI, Vol. 1652, 1992, 23-30.

[MiK] D.L. Milgram and D.J. Kahl, *Recursive region extraction.* Computer Graphics and Image Processing, 9, 82-88, 1979.

[Mi] D.L. Milgram, *Region extraction using convergent evidence.* Computer Graphics and Image Processing, 11, 1-12, 1979.

[MMP] J. Marroquin, S. Mitter and T. Poggio, *Probabilistic solution of ill-posed problems in computational vision.* J. American Statistical Association, March 1987.

[Mont] G.H. Montaneri, *On the optimal detection of curves in noisy pictures.* Comm. Assoc. Comput. Mach. 14, 335-345, 1971.

[MueA] J.L. Muerle and D.C. Allen, *Experimental evaluation of techniques for automatic segmentation of objects in a complex scene.* In Pictorial Pattern Recognition, G.C.Cheng and al. eds., pp 3-13, Thompson, Washington, 1968.

[NahLM] N.E. Nahi and S. Lopez Mora, *Estimation-detection of object boundaries in noisy images.* IEEE Trans. Automat. Control, 23, 834-846, 1978.

[NarLR] K.A. Narayanan and D.P. O'Leary and A. Rosenfeld, *Image smoothing and segmentation by cost minimization.* IEEE Trans. Syst. Man, Cybern. 12(1), 91-96, 1982.

[NiM] M. Nitzberg and D. Mumford, *The 2.1-D sketch.* Preprint, Department of Mathematics, Harvard University Cambridge MA.

[Nor] N.K. Nordström, *Variational edge detection.* PhD dissertation, Department of electrical engineering, University of California at Berkeley, 1990.

[OhPR] R. Ohlander, K. Price and P.R. Reddy, *Picture segmentation using a recursive region splitting method.* Comput. Graphics Image Process. 8, 313-333, 1978.

[PanR] D. P. Panda and A. Rosenfeld, *Image segmentation by pixel classification in (gray level, edge value) space.* IEEE Trans. Comput. C-27, 875-879, 1978.

[PapJ] N.T. Pappas and N.S. Jayant, *An adaptive clustering algorithm for image segmentation.* Proc. of ICCV'88, IEEE n° 883, 310-315, 1988.

[ParM] B. Parvin and G. Medioni, *Segmentation of range images into planar surfaces by split and merge.* Proc. of Computer Vision and Pattern Recognition Conf., IEEE, 1986.

[Pav1] T. Pavlidis, *Segmentation of pictures and maps through functional approximation.* Comp.Gr. and Im.Proc. 1, 360-372, 1972.

[PavL] T. Pavlidis and Y.T. Liow, *Integrating Region Growing and Edge Detection.* Proc. of the IEEE Conf. Comp. Vision and Patt. Recognition 1988.

[Pel] S. Peleg, *Classification by discrete optimization.* Computer Vis., Graphics, Image Processing, 25(1), 122-130, 1984.

[Pen1] A. Pentland, *Fractal-based description of natural scenes.* IEEE Trans. PAMI, 6(6), 661-674, 1984.

[Pen2] A.P. Pentland, *Automatic extraction of deformable part models.* International Journal of Computer Vision, 4, 107-126, 1990.

[Per] W.A. Perkins, *Area segmentation of images using edge s.* IEEE PAMI 2(1), January 1980.

[PonSWH] T.C. Pong, L.G. Shapiro, L.T.Watson and R.M. Haralick, *Experiments in segmentation using a facet model region grower.* Computer Vision and Graphics Image Processing, 25, 1984.

[Pra] J.M. Prager, *Extracting and labeling boundary segments in natural scenes.*

IEEE PAMI 2(1), 16-27, January 1980.

[PreM] J.S.M. Prewitt and M.L. Mendelsohn, *The analysis of cell images.*
Ann. N.Y. Acad. Sci. 128, 1035-1053, 1966.

[RisA] E. Riseman and M. Arbib, *Computational techniques in the visual segmentation of static scenes.* Computer Graphics and Image Processing,
6, 221-276, 1977.

[ROF] L. Rudin, S. Osher and E. Fatemi, *Nonlinear Total Variation Based
Noise Removal Algorithms.* Proc. Modélisations matématiques pour le
traitement d'images, INRIA, 149-179 and Physica D., Proceedings of
11th Conf. on Experimental Mathematics, Sept 1992.

[RosD] A. Rosenfeld and L.S. Davis, *Image segmentation and image models.*
Proc. IEEE, vol 67(5), 764-772, May 1979.

[RosHZ] A. Rosenfeld, R.A. Hummel and S.W. Zucker, *Scene labeling by relaxation operations.* IEEE Trans. Syst., Man, Cybern., SMC-6, 420-433,
1976.

[RosT] A. Rosenfeld and M. Thurston, *Edge and curve detection for visual
scene analysis.* IEEE Trans. Comput. 20, 562-569, 1971.

[RotL] G. Roth and M.D. Levine, *Segmentation of geometric signals using
robust fitting.* IEEE Proc. 10th Int. Conf. on Pattern Recognition (I),
826-831, Atlantic City 1990.

[RuKo] L. Rudin and G. Koepfler, *Nonlinear variational image clutter removal
via pyramidal domain decomposition,* preprint, 1993.

[ScDR] B.J. Schachter, L.S. Davis and A. Rosenfeld, *Some experiments in image segmentation by clustering of local feature values.* Pattern Recognition 11, 19-28, 1978.

[Sha'U] A. Sha'ashua and S. Ullman, *Structural Saliency: the detection of glob-
ally salient structures using a locally connected network.* Proc. of the
IEEE 2nd International Conference on Computer Vision, Tampa, 1988.

[ShanDG] K.F. Shanmugam, F.M. Dickey and J.A. Green, *An optimal frequency
domain filter for edge detection in digital pictures.* IEEE PAMI 1, 37-49,
1979.

[Sim-JC] J.C. Simon, *Some current topics in clustering in relation with pattern
recognition.* Proc. 4th Int. Joint Conf. on Pattern Recognition, 19-29,
1978.

[SnS] P.H.A. Sneath and R.R. Sokal, *Numerical taxonomy.* San Francisco
CA, Freeman 1973.

[Te1] D. Terzopoulos, *Multilevel computational processes for visual surface
reconstruction.* Computer Vision, Graphics, and Image Processing, 24,
52-95, 1983.

[Te2] D. Terzopoulos, *Computing visible-surface representations.* A.I. Memo
800, MIT Artificial Intelligence Lab., Cambridge MA, March 1985.

[Te3]   D. Terzopoulos, *Regularization of inverse visual problems involving discontinuities.* IEEE PAMI-8,4, July 1986.

[TorP]   V. Torre and T.A. Poggio, *On edge detection.* IEEE PAMI 8(2), 147-163, 1986.

[Vin]   L. Vinet, *Segmentation et mise en correspondance de régions de paires d'images stéréoscopiques.* Thèse Univ. Paris IX Dauphine, juillet 1991.

[War]   J.H. Ward, *Hierarchical grouping to optimize an objective function.* J. Amer. Stat. Ass. 58, 236-245, 1963.

[WeiB]   R. Weiss and M. Boldt, *Geometric grouping applied to straight lines.* Proc. of the IEEE Conf. on Computer Vision and Pattern Recognition, 1986.

[WestHLP]   T. Westman, D. Harwood, T. Laitinen and M. Pietikinen, *Color segmentation by hierarchical connected components analysis with image enhancement by symmetric neighbourhood filters.* IEEE Proc. 10th Int. Conf. on Pattern Recognition (I), 796-802, Atlantic City 1990.

[Yui]   A.L. Yuille, *Generalized deformable models, statistical physics and matching problems.* Review, Neural Computation 2, 1-24 1990.

[ZuDDI]   S.W. Zucker, C. David, A. Dobbins and L. Iverson, *The organization of curve detection: Coarse tangent fields and fine spline coverings.* Proc. 2nd Int. Conf. Computer Vision, 568-577, Tampa, Florida 1988.

I-C) Scale space theory (Part I).

[AGLM1]   L. Alvarez, F. Guichard, P.L. Lions and J.M. Morel, *Axioms and fundamental equations of image processing* Report 9216, 1992 CEREMADE. Université Paris Dauphine . Arch. for Rat. Mech. 16, IX, 200-257, 1993

[AGLM2]   L. Alvarez, F. Guichard, P.L. Lions and J.M. Morel, *Axiomatisation et nouveaux opérateurs de la morphologie mathématique,* C.R. Acad. Sci. Paris, t.315, Série I 265–268, 1992.

[AlMa]   L. Alvarez and L. Mazorra, *Signal and Image Restoration by using shock filters and anisotropic diffusion.* Preprint of Dep. de Inf. U.L.P.G.C. ref:0192, to appear in SIAM journal on numerical analysis, 1992.

[ASB]   H. Asada and M. Brady, *The curvature primal sketch.* IEEE Trans. PAMI 8(1), 2-14, 1986.

[BabWBD]   J. Babaud, A. Witkin, A. Baudin and R. Duda, *Uniqueness of the gaussian kernel for scale-space filtering.* IEEE Trans. PAMI 8, January 1986.

[BrMa]   W. Brockett and P. Maragos, *Evolution Equations for continuous-scale morphology,* ICASSP, San Francisco 23–26, 1992.

[BSS]   A.M. Bruckstein, G. Sapiro and D. Shaked, *Affine invariant evolutions of planar polygons,* preprint, 1992.

[CEGLM] T. Cohignac, F. Eve, F. Guichard, C. Lopez, J.M. Morel, *Numerical Analysis of the fundamental equation of image processing*, submitted to Int. J. of Comp. Vision, 1993.

[Dia] J.I. Diaz, *A nonlinear parabolic equation arising in image processing*. Extracta Matematicae, Universidad de Extremadura 1990.

[Fau] O. Faugeras, *A few steps toward a projective scale space analysis*, C.R. Acad. Sci. Paris, to appear, 1993.

[Flo] L.M.J. Florack, *The syntactical structure of scalar images*. PhD thesis, University of Utrecht, The Netherlands, 1993.

[FMZ] D. Forsyth, J.L. Mundy, A. Zisserman, al., *Invariant descriptors for 3-D object recognition and Pose*, IEEE Transaction of pattern analysis and machine intelligence, **13**, No.10, 1991.

[FRKV] L. Florack, B. ter Haar Romeny, J.J. Koenderink and M. Viergever, *General Intensity transformations and Second Order Invariants*, Proceedings of the 7th Scandinavian Conference on Image Analysis. Aalborg13–16, 1991.

[FRKV] L. Florack, B. ter Haar Romeny, J.J. Koenderink and M. Viergever, *Scale and the differential structure of images* Image and vision computing **10** No. 6, 1992.

[Gra] R.E. Graham, *Snow removal: A noise-stripping process for TV signals*. IRE Trans. Inf. Theory IT-9, 129-144, 1962.

[Gui] F. Guichard, *Multiscale Analysis of Movies*, Proceedings of the eighth workshop on image and multidimensional signal processing, IEEE, September 8-10, Cannes. 236–237, 1993.

[HumM] R. Hummel and R. Moniot, *Reconstruction from zero crossings in scale-space*. IEEE Trans. on Acoustic Speech and Signal Processing. December 1989.

[Hum] R. Hummel, *Representations based on zero-crossing in scale-space*, Proc. IEEE Co

[HornW] B.K.P. Horn and E.J. Weldon. *Filtering closed curves*. IEEE Trans. PAMI 8(5), 665-668, 1986.

[Kim] B.B. Kimia, *Toward a computational theory of shape, Ph.D. Dissertation*, Department of Electrical Engineering, McGill University, Montreal, Canada, 1990.

[KoDo1] J.J. Koenderink and A.J. van Doorn, *Dynamic shape*, Biol. Cyber. **53**, 383–396, 1986.

[KoDo2] J.J. Koenderink and A.J. van Doorn, *Representation of Local Geometry in the Visual System*, Biol. Cyb. **55**, 367–375, 1987.

[KTZ] B.B. Kimia, A. Tannenbaum, and S.W. Zucker, *On the evolution of curves via a function of curvature, 1: the classical case*, J. of Math. Analysis and Applications **163**, No 2, 1992.

[LevZR] A. Lev, S.W. Zucker and A. Rosenfeld. *Iterative Enhancement of noisy images*. IEEE SMC-7, 435-442, 1977.

[Lin] T. Lindeberg, *Scale-space for discrete signal*, IEEE Trans. Pattern Anal. and Machine Intell. **12**, 234–254, 1990.

[Log] D. Logan, *Information in the zero-crossings of band pass signals*. Bell Systems Tech. Journal, 56, 510-, 1977.

[Low2] D.G. Lowe, *Organization of smooth image curves at multiple scales*. International Journal of Computer Vision, 3, 119-130, 1989.

[LSW] Y. Lamdan, J.T. Schwartz and H.J. Wolfson, *Object recognition by affine invariant matching* In Proc. CVPR 1988.

[MaliP1] P. Perona and J. Malik, *A scale space and edge detection using anisotropic diffusion*. Proc. IEEE Computer Soc. Workshop on Computer Vision, 1987.

[MaMo1] A. Mackworth and F.Mokhtarian, *Scale-Based description and recognition of planar curves and two-dimensional shapes*, IEEE Transactions on Pattern analysis and machine intelligence **8**, No. 1, 1986.

[MaMo2] A. Mackworth and F.Mokhtarian, *A theory of multiscale, curvature-based shape representation for planar curves*, IEEE Trans. Pattern Anal. Machine Intell. **14** pp 789– 805, 1992.

[Mara1] P. Maragos, *Tutorial on advances in morphological image processing and analysis* optical engineering **26** No. 7, 1987.

[Marr1] D.Marr, *Early processing of visual information*. Philos. Trans. Roy. Soc. London Ser. B, 275, 483-524, 1976.

[NiS] M. Nitzberg and T. Shiota, *Nonlinear image smoothing with edge and corner enhancement*. Technical report n° 90-2, Harvard university, Cambridge, MA 01238, 1990.

[OshR] S. Osher and L. Rudin, *Feature-oriented image enhancement using shock filters*. SIAM J. on Numerical Analysis, 27, 919-940, 1990.

[OshR] S. Osher and L. Rudin, *Shocks and other nonlinear filtering applied to image processing* Proc. SPIE Appl. of Digital Image Processing XIV, Vol 1567, 414-430, 1991.

[PFMVG] E.J. Pauwels, P. Fiddelaers, T. Moons and L.J. Van Gool, *An extended class of scale-invariant and recursive scale space filters*, preprint, submitted to PAMI.

[PeMa] P. Perona and J. Malik, *A scale space and edge detection using anisotropic diffusion*, Proc. IEEE Computer Soc. Workshop on Computer Vision, 1987.

[PogTK] T. Poggio, V. Torre and C. Koch, *Computational vision and regularization theory*. Nature, 317, 1985.

[PWO] C.B. Price, P. Wanback and A. Oosterlinck, *Application of reaction-diffusion equations to image processing*. Proc. 3rd Internat. Conf. Image Processing and its Applications, 49-53, IEEE Publ., London, 1989.

[Ru] L.I. Rudin, *Images, numerical analysis of singularities and shock filters.* Ph.D dissertation, Caltech, Pasadena , California n5250:TR, 1987.

[RuOs] L.I. Rudin and S. Osher, *Image processing makes its mark in court.* Interview in SIAM news, 26, 8, Dec.1993.

[SaTa1] G. Sapiro and A. Tannenbaum, *On affine plane curve evolution*, Department of electrical engineering. Technion Israel Institute of Technology. Haifa. Israel. EE Pub 821 1992 submitted.

[SaTa2] G. Sapiro and A. Tannenbaum, *Affine shortening of non-convex plane curves*, Department of electrical engineering. Technion Israel Institute of Technology. Haifa. Israel. EE Pub 845, 1992 submitted.

[Weic] J.Weickert, *Anisotropic diffusion filters for image processing based quality control.* Proc. 7th Conf. on Math. in Industry, A. Fasano (Ed.), Montecatini Terme, 1993.

[Wi2] A.P. Witkin, *Scale-space filtering.* Proc. of IJCAI, Karlsruhe 1983, 1019-1021.

[Yui] A. Yuille, *The creation of structure in dynamic shape*, Proc. of the second international conference on computer vision, Tampa 685–689, 1988.

[YuiPo] A.L. Yuille and T.A. Poggio, *Scaling theorems for zero-crossings.* IEEE Trans. Pattern Anal. Machine Intell., 8, 15-25, 1986.

I-D) Scale space theory (Part I).

[Ang] S. Angenent, *Parabolic equations for curves on surfaces I, II* University of Wisconsin-Madison Technical Summary Reports **19, 24**, 1989.

[BaGe] G. Barles and C. Georgelin, *A simple proof of convergence for an approximation scheme for computing motions by mean curvature*, preprint, 1992.

[Bar] G. Barles, *Remarks on a flame propagation model*, Technical Report No 464 INRIA Rapports de Recherche, 1985.

[BaSo] G. Barles and P.E. Souganidis, *Convergence of approximation schemes for fully nonlinear second order equation*, Asymp. Anal., 1993 to appear.

[CDK] F. Catté, F. Dibos and G. Koepfler, *A morphological approach of mean curvature motion* Report 9310, CEREMADE, Univ. Paris Dauphine, 1993.

[ChGG] Y.G. Chen, Y. Giga and S. Goto, *Uniqueness and existence of viscosity solutions of generalized mean curvature flow equations,* preprint, Hokkaido University, 1989.

[CIL] M.G. Crandall, H. Ishii and P.-L. Lions, *User's guide to viscosity solution of second order partial differential equation*, CEREMADE, preprint, 1991.

[EvS] L.C. Evans and J. Spruck, *Motion of level sets by mean curvature I*, preprint.

[GaHa] M. Gage and R.S. Hamilton, *The heat equation shrinking convex plane curves*, J. Differential Geometry **23**, 69–96, 1986.

[GiGIS] Y. Giga, S. Goto, H. Ishii and M.H. Sato. *Comparison principle and convexity preserving properties for singular degenerate parabolic equations on unbounded domains*. Preprint Hokkaido University, 1990.

[GLM] F. Guichard, J.M. Lasry and J.M Morel, *A monotone consistent theoretical scheme for the fundamental equation of image processing*, preprint, 1993.

[Gray] M. Grayson, *The heat equation shrinks embedded plane curves to round points*, J. Differential Geometry **26**, 285–314, 1987.

[HolN] K. Hollig and J.A. Nohel, *A diffusion equation with a nonmonotone constitutive function*. Proc. on systems of nonlinear PDEs, John Ball ed., Reidel, p. 409-422, 1983.

[Mas] P. Mascarenhas, *Diffusion generated motion by mean curvature*, preprint, 1992.

[MBO] B. Merriman, J. Bence and S. Osher, *Diffusion Generated motion by Mean Curvature*, CAM Report 92-18. Depart. of Mathematics, University of California, Los Angeles CA 90024.1555, 1992.

[OshS] S. Osher and J. Sethian, *Fronts propagating with curvature dependant speed: algorithms based on the Hamilton-Jacobi formulation*. J. Comp. Physics, 79, 12-49, 1988.

[OsSe] S. Osher and J. Sethian, *Fronts propagating with curvature dependent speed: algorithms based on the Hamilton-Jacobi formulation*, J. Comp. Physics **79**, 12–49, 1988.

[Son] M. Soner, *Motion of a set by curvature of its mean boundary*. Preprint

I-E) Monographs in Image Processing.

[AhS] N. Ahuja and B. Schachter, *Pattern Models*. Wiley, New York 1983.

[BalB] D.H. Ballard and C.M. Brown, *Computer Vision*. Englewood Cliffs, NJ, Prentice Hall 1982.

[Bod] M. Boden, *Computer models of mind*. Cambridge University Press 1987.

[Horn] B. K. P. Horn, *Robot Vision*. MIT Press 1987.

[Koel] J.J. Koenderink, Solid Shape MIT Press, Cambridge, MA, 1990.

[Low1] D.G. Lowe, *Perceptual Organization and Visual Recognition*. Kluwe Academic Publishers, 1985.

[Marr3] D. Marr, *Vision*. Freeman and Co. 1982.

[Math] G. Matheron, *Random Sets and Integral Geometry*, John Wiley, N.Y., 1975

[Pav2] T. Pavlidis, *Structural Pattern Recognition*. Springer, New York 1977.

[Roc] I. Rock, *The logic of perception.* Cambridge MA, MIT Press, 1983.

[RosK] A. Rosenfeld and Kak A.C. *Digital picture processing, Computer Science and Applied Mathematics.* Academic Press, 1982.

[Ser] J. Serra, *Image analysis and mathematical morphology,* Vol 1, Academic Press, 1982.

[TanK] S. Tanimoto and A. Klinger, *Structural Computer Vision.* New York: Academic, 1980.

[TouG] J.T. Tou and R.C. Gonzalez, *Pattern Recognition Principles.* Reading, MA: Addison Wesley 1974.

I-F) Articles on texture analysis and segmentation.

[Bec1] J. Beck, *Similarity grouping and peripheral discriminability under uncertainty.* American Journal of Psychology, 85, 1-19, 1972.

[Bec2] J. Beck, *Organization and representation in perception.* In: *Textural segmentation,* Erlbaum, Hillsdale, NJ, 1982.

[BecPR] J. Beck, K. Prazdny and A. Rosenfeld, *A theory of textural segmentation.* Human and Machine Vision, J.Beck, B. Hope and A. Rosenfeld eds., New York Academic Press, 1-38,1980.

[BecSI] J. Beck, A. Sutter and R. Ivry, *Spatial frequency channels and perceptual grouping in texture segregation.* Computer Vision, Graphics and Image Processing, 37, 299-325, 1987.

[BenB] P. Bennett and M.S. Banks, *Sensitivity loss in odd-symmetric mechanisms and phase anomalies in peripheral vision.* Nature, 326, 873-876, 1987.

[BenH] B.M. Bennett and D.D. Hoffman. In: *Image Understanding.* 1985-1986, W. Richards and S. Ullman eds, Ablex Publishing Corporation, Norwood, NJ, 1987.

[BergenA] J.R. Bergen and E.H. Adelson, *Early vision and texture perception.* Nature, 333, 363-364, 26 May 1988.

[BergenJ1] J.R. Bergen and B. Julesz, *Textons, the fundamental elements of preattentive vision and perception of textures.* Bell System Techical Journal, 62(6), 1619-1645, 1983.

[BergenJ2] J. Bergen and B. Julesz, *Rapid discrimination of visual patterns.* IEEE Trans. on Systems, Man, and Cybernetics, 13 (5), 1983.

[BovCG] A.C. Bovik, M. Clark and W. S. Geisler. *Computational texture analysis using localized spatial filtering.* IEEE, 1987.

[Bro] Ph. Brodatz, *Textures for artists and designers.* Dover Publications Inc. NY, 1966.

[Cae] T. Caelli, *Three processing characteristics of visual texture segmentation.* Spatial Vision, 1(1), 19-30, 1985.

[CLM] T. Cohignac, C. Lopez, and J.M. Morel, *Multiscale analysis of shapes, images and textures,* Proceedings of the eighth workshop on image and multidimensional signal processing, IEEE, September 8-10, Cannes. 142–143, 1993.

[CogJ]   J.M. Coggins and A.K. Jain, *A spatial filtering approach to texture analysis.* Pattern Recog. Lett., 3, 195-203, 1985.

[ConnH]   R.W. Conners and C.A. Harlow, *A theoretical comparison of texture algorithms.* IEEE Trans. PAMI 2(3), 204-222, 1980.

[D'AJ]   F. D'Astous and M.E. Jernigan, *Texture discrimination based on detailed measures of the power spectrum.* Proc. 7th Internat. Conf. Pattern Recog., 83-86, 1984.

[DavCA]   L.S. Davis, N. Clearman and J.K. Aggarwal, *An empirical evaluation of generalized co-occurrence matrices.* IEEE Trans. PAMI 3(2) 214-221, 1981.

[En]   J. Enns, *Seeing textons in context.* Perception and Psychophysics, 39 (2), 143-147, 1986.

[FiN]   D. Field and J. Nachmias, *Phase reversal discrimination.* Journal of Vision Research, 24, 333-340, 1984.

[GraBS]   M. Graham, J. Beck and A. Sutter, *Two nonlinearities in texture segregation.* Proceedings of the ARVO conference, p161, 1989.

[GurB]   R. Gurnsey and R. Browse, *Micropattern properties and presentation conditions influencing visual texture discrimination.* Perception and Psychophysics, 41 (3), 239-252, 1987.

[Har1]   R.M. Haralick, *Statistical and structural approaches to textures.* Proceedings of the IEEE, 67 (5): 786-804, 1979.

[Ju1]   B. Julesz, *Spatial nonlinearities in the instantaneous perception of textures with identical power spectra.* Phil. Trans. R. Soc. Lond., B290, 83-94, 1980.

[Ju2]   B. Julesz, *Textons, the elements of texture perception, and their interactions.* Nature, 290, 12 March 1981.

[Ju3]   B. Julesz, *Texton gradients: the texton theory revisited.* Biological Cybernetics, 54: 245-251, 1986.

[JuGV]   B. Julesz, E.N. Gilbert and J. D. Victor, *Visual discrimination of textures with identical third order statistics.* Biological Cybernetics, 31, 137-140, 1978.

[JuK]   B. Julesz and B. Kr}ose, *Features and spatial filters.* Nature, 333, 302-303, May 1988.

[KjD]   B.P. Kjell and C.R. Dyer, *Edge separation and orientation texture measures.* Proc. CVPR-85, 306-311, 1985.

[Kr]   B.J. Kröse, *Local structure analyzers as determinants of preattentive pattern discrimination.* Biological Cybernetics, 55, 289-298, 1987.

[LeuW]   J.-G. Leu and W. Wee, *Detecting the spatial structure of natural textures based on shape analysis.* Comput. Vision, Graphics, Image Process. 31, 67-88, 1985.

[LoMo] C. Lopez, J.M. Morel, *Axiomatization of shape analysis and application to texture hyperdiscrimination*. Proceedings of the Trento Conference on Surface Tension and Movement by Mean Curvature, De Gruyter, Berlin, 1992.

[MaleBF] J.T. Maleson, C.M. Brown and J.A. Feldman, *Understanding natural texture*. Proc. DARPA Image Understanding Workshop, 19-27, October 1977.

[MaliP] J. Malik and P. Perona, *Preattentive texture discrimination with early vision mechanisms*. Journ. of the Opt. Society. of America. A, 923-932, vol. 7, n° 5, 1991.

[Marr2] D. Marr, *Analysing natural images: a computational theory of texture vision*. Cold Spring Harbor Symposium on Quantitative Biology, XL, 647-662, 1976.

[MatSN] T. Matsuyama, K. Saburi and M. Nagao, *A structural analyser for regularly arranged textures*. Comput. Graphics Image Process. 18, 259-278, 1982.

[Mue] J.L. Muerle, *Some thoughts on texture discrimination by computer*. In: B.S. Lipkin and A. Rosenfeld (eds), Picture Processing and Psychopictorics, Academic Press, New York, 371-379, 1970.

[Not] H.C. Nothdurft, *Orientation sensitivity and texture segmentation*. Vision Res., 25, 551- 560, 1985.

[PelNHA] S. Peleg, J. Naor, R. Hartley and D. Avnir, *Multiple resolution texture analysis and classification*. IEEE Trans. PAMI 6(4), 518-523, 1984.

[Rea] T.C. Rearick, *A texture analysis algorithm inspired by a theory of preattentive vision*. Proc. IEEE Conf. Comput. Vision Pattern Recog., 312-317, San Francisco 1985.

[RenHC] I. Rentschler, M. Hubner and T. Caelli, *On the discrimination of compound Gabor signals and textures*. Vision Research, 28 (2), 279-291, 1988.

[RuS] B. Rubenstein and D. Sagi, *Texture variability across the orientation spectrum can yield asymmetry in texture discrimination*. European conference on visual perception, 1989.

[SuBG] A. Sutter, J. Beck and N. Graham, *Contrast and spatial variables in texture segregation : testing a simple spatial frequency channels model*. Center for Automation Research, Computer Vision Lab., CAR-TR-381, CS-TR-2091, DCR-86-03723, Univ. of Maryland, College Park, 1988.

[TamMY] H. Tamura, S. Mori and T. Yamawaki, *Textural features corresponding to visual perception*. IEEE Trans. Syst., Man and Cybern., 8(6), 460-473, 1978.

[TomST] F. Tomita, Y. Shirai and S. Tsuji, *Description of textures by a structural analysis*. Proc. 6th Intern. Joint Conf. Artif. Intell., 884-889, Tokyo 1979.

[Tr] A. Treisman, *Preattentive processing in vision.* Comput. Vision, Graphics, Image Process., 31(2), 156-177, 1985.

[TsT] S. Tsuji and F. Tomita, *A structural analyser for a class of textures.* Comput. Graphics Image Process., 2, 216-231, 1973.

[Tu] M.R. Turner, *Texture discrimination by Gabor functions.* Biol. Cybern. 55, 71-82, 1986.

[VilNP] F. Vilnrotter, R. Nevatia and K. Price, *Structural analysis of natural textures.* IEEE Trans. PAMI 8, 76-89, 1986.

[Vis] R. Vistnes, *Texture models and image measures for texture discrimination.* International Journal of Computer Vision, 3, 313-336, 1989.

[VoP1] H. Voorhees and T. Poggio, *Detecting textons and texture boundaries in natural images.* Proc. Int. Conf. Comp. Vision, IEEE, 250-258, 1987.

[VoP2] H. Voorhees and T. Poggio, *Computing texture boundaries from images.* Nature, 333, 364-367, 1988.

[WermL] D. Wermser and C.-E. Liedtke, *Texture analysis using a model of the visual system.* Proc. 6th Intern. Conf. Pattern Recog., 1078-1080, Munich 1982.

[Wil] A.P. Witkin, *Recovering surface shape and orientation from texture.* Artificial Intelligence 17, 17-45, August 1981.

I-G) Wavelets, relation to image processing and scale space.

[CohDF] A. Cohen, I. Daubechies and J.C. Feauveau, *Biorthogonal bases of compactly supported wavelets.* Preprint Bell Labs, 1989.

[CohF] A. Cohen and J. Froment, *A mathematical and practical analysis of image compression by wavelet transform.* To appear.

[Coh] A. Cohen, *Ondelettes, analyses multirésolutions et filtres miroir en quadrature.* Ann. Inst. Henri Poincaré, Analyse non linéaire, Vol. 7, n°5, p.439-459, 1990.

[Daub1] I. Daubechies, *Orthonormal bases of compactly supported wavelets.* Comm. Pure and Applied Math., vol. 41, p.909-996, 1988.

[Daub2] I. Daubechies, *Orthonormal bases of compactly supported wavelets.* Part II, II, Preprint Bell Labs, 1989.

[DaubJJ] I. Daubechies, S. Jaffard and J.L. Journé, *A simple Wilson orthonormal basis with exponential decay.* Preprint Bell Labs, 1989.

[Daug1] J.G. Daugman, *Two-dimensional spectral analysis of cortical receptive field profiles.* Vision Res., 20, 847-856, 1980.

[Daug2] J.G. Daugman, *Six formal properties of two-dimensional anisotropic visual filters: structure principles and frequency/orientation selectivity.* IEEE Trans. Syst., Man, Cybern., 13(5), 1984.

[Daug3] J.G. Daugman, *Uncertainty relation for resolution in space, special frequency and orientation optimized by two-dimensional visual cortical filters.* J. Opt. Soc. Am., A, 2, 1160-1169, 1985.

[Dau] I. Daubechies, *Ten lectures on wavelets*, SIAM, 1992.

[DVL] R. A. DeVore and B.J. Lucier, *Fast wavelet techniques for near-optimal image processing*. IEEE Military Communications Conference, 1992.

[EsG] D. Esteban and C. Galand, *Application of QMF to split-band voice coding schemes*. Proc. of Int. Conf. Acoust. Speech, Signal Processing, Harford, 1977.

[FroM] J. Froment and J.M. Morel, *Analyse multiéchelle, vision stéréo et ondelettes*. Exposé n° 5 du Séminaire d'Orsay "Les ondelettes en 89", ed. P.G. Lemarié, Springer Verlag, 1990.

[Fro] J. Froment, *Traitement d'images et applications de la transformée en ondelettes*. Thèse Univ. Paris IX Dauphine, 1990

[Gab] D. Gabor, *Theory of communication*. J. Inst. Elect. Engr. 93, 429-457, 1946.

[GroM] A. Grossmann and J. Morlet, *Decompositions of Hardy functions into square integrable wavelets of constant shape*. SIAM Journal Math. Anal. 15, 723-736, 1984.

[Haa] J.A. Haar, *Zur Theorie der orthogonalen Funktionensysteme*. Math. Ann. 69, 331-371, 1910

[Hum] R.A. Hummel, *Feature detection using basic functions*. Computer Graphics and Image Processing, 9, 40-55, 1979.

[IkU] A. Ikonomopoloulos and M. Unser, *A directional filtering approach to texture discrimination*. Proc. 7th Internat. Conf. Pattern Recog., 87-89, 1984.

[LemM] P.G. Lemarié et Y. Meyer, *Ondelettes et bases hilbertiennes*. Revista Matematica Iberoamericana, Vol.2, 1-18, 1986.

[MallZ] S.G. Mallat and S. Zhong, *Complete signal representation with multiscale edges*. Technical report n° 483, Robotics report n° 219, NYU, Computer Science Division, Courant Institute, December1989.

[Mall] S. Mallat, *A theory for multiresolution signal decomposition: the wavelet representation*. IEEE Trans. on Pattern Analysis and Machine Intelligence, 674-693, July 1989.

[Mey] Y. Meyer, *Ondelettes et algorithmes concurrents*, Hermann, 1992

[Me] Y. Meyer, *Ondelettes, fonctions splines et analyses graduées*. Rapport CEREMADE n°8703, 1987.

[PolR] D.A. Pollen and S.F. Ronner, *Visual cortical neurones as localized spatial frequency filters*. IEEE Trans. Syst., Man, Cybern. 13, 907-916, 1983.

I-H) Related topics in psychophysics, neurobiology and gestalt theory.

[HePB] R. von der Heydt, E. Peterhans and G. Baumgartner, *Illusory contours and cortical neuron responses*. Science 224, 1260-1262, 1984.

[HubW]   D.H. Hubel and T.N. Wiesel, *Receptive fields, binocular interaction and functional architecture in the cat's visual cortex.* J. Physiol. 160, 106-154, 1962.

[Kani]   G. Kanizsa, *Organization in Vision.* New York: Praeger 1979.

[Koe2]   J.J. Koenderink, *The brain a geometry engine* Psychol. Res.**52**, 122-127, 1990.

[Koen]   J.J. Koenderink, *The structure of images.* Biol. Cybern. 50, 363-370, 1984.

[Koh]   R. Köhler, *A segmentation system based on thresholding.* Computer Graphics and Image Processing, 15, 319-338, 1981.

[Köh]   W. Köhler, *Psychologie de la forme.* Idées nrf-Gallimard, 1964.

[Marc]   S. Marcelja, *Mathematical description of the response of simple cortical cells.* J. Opt. Soc. Am. 70, 1297-1300, 1980.

[Mum]   D. Mumford, *On the computational architecture of the neocortex. I the role of the thalamo-cortical loop.* Biol. Cybern., 1991.

[NakSS]   K. Nakayama, S. Shimojo and G. Silverman, *Stereoscopic depth: its relation to image segmentation, grouping, and the recognition of occluded objects.* Manuscript submitted to Perception, October 1988.

[WebDV]   M.A. Webster and R.L. De Valois, *Relationship between spatial-frequency and orientation tuning of straite-cortex cells.* J. Opt. Soc. Am. 2, 1124-1132, 1985.

[Wert]   M. Wertheimer, *Untersuchungen zur Lehre von der Gestalt II.* Psychologische Forschung, 4, 301-350, 1923.

II-A) References on geometric measure theory and rectifiability.

[BarD]   M.F. Barnsley and S. Demko, *Iterated function schemes and the global construction of Fractals.* Proc. R. Soc. London. Ser. A 399, 243-275, 1985.

[Besi1]   A.S. Besicovitch, *On the fundamental geometrical properties of linearly measurable plane sets of points.* Math.Ann. vol. 98, 422-464, 1928.

[Besi2]   A.S. Besicovitch, *On the fundamental geometrical properties of linearly measurable plane sets of points II.* Math.Ann. vol. 115, 296-329, 1938.

[Besi3]   A.S. Besicovitch, *On the fundamental geometrical properties of linearly measurable plane sets of points III.* Math.Ann. vol. 116, 349-357, 1939.

[Cara]   C. Carathéodory, *Über das lineare Mass von Punktmengen. EineVerallgemeinerung des Längenbegriffs.* Nach. Ges. Wiss. Göttingen, 404-426, 1914.

[DaSe1]   G. David and S. Semmes, *Analysis of and on uniformly rectifiable sets.* AMS Mathematical Surveys and Monographs, 38, 1993.

[DaSe2]   G. David and S. Semmes, *Singular integrals and rectifiable sets in $\mathbb{R}^N$. Au dela des graphes lipschitziens.* Astérisque n° 193, 1991.

[Fed1] H. Federer, *The $((\varphi, k)$ rectifiable subsets of n-space*, Trans. Amer. Math. Soc. 62, 114-192, 1947.

[Gi] J. Gillis, *On the projection of irregular linearly measurable plane sets of points.* Fund. Math., 26, 229-233, 1934.

[Haus] F. Hausdorff, *Dimension und äusseres Mass.* Mathematische Annalen, 79, 157- 179, 1919.

[Jon] P.W. Jones, *Rectifiable sets and the travelling salesman problem*, Invent. Math. 102, 1-15, 1990.

[Mars1] J.M. Marstrand, *The (Φ, s)-regular subsets of n-space.* Trans. A.M.S. 113, 369-392, 1964.

[Mars] J.M. Marstrand, *Hausdorff two-dimensional measure in 3-space*, Proc. London Math. Soc. 11, 91-108, 1961.

[Matt1] P. Mattila, *Hausdorff m-regular and rectifiable sets in n-space.* Trans. A.M.S. 205, 263-274, 1975.

[Preis] D. Preiss, *Geometry of measures in $\mathbb{R}^N$: distribution, rectifiability, and densities.* Annals of Mathematics, 125, 537-643, 1987.

[Vol] A.I. Vol'pert, *The space BV and quasilinear equations*, Math. USSR, Sbornik, 2, 225-267, 1967.

II-B) Monographs on mathematics and geometric measure theory.

[Ale] P.S. Aleksandrov, *Combinatorial Topology.* Vol. 1, chap. 2, Graylock Press NY, 1956.

[Bre] H. Brezis, *Analyse Fonctionnelle, théorie et applications.* Masson, Paris, 1983.

[EvGa] L.C. Evans and F. Gariepy, *Lecture Notes on Measure Theory and Fine Properties of Functions.*, CRC Press, 1992.

[Egg] H.G. Eggleston, *Convexity.* Cambridge University Press, 1958.

[Fal1] K.J. Falconer, *The Geometry of Fractal Sets.* Cambridge University Press, 1985.

[Fal2] K.J. Falconer, *Fractal Geometry: Mathematical Foundations and Applications.* J.Wiley and Sons,1990.

[Fed] H. Federer, *Geometric Measure Theory.* Springer-Verlag,1969.

[GilT] D. Gilbarg and N.S. Trudinger, *Elliptic partial differential equations of second order*, second edition, Springer-Verlag 1983.

[Giu] E. Giusti, *Minimal Surfaces and Functions of Bounded Variation.* Birkhäuser 1984.

[KiSt] D. Kinderlehrer and G. Stampacchia, *Variational inequalities and applications*, Academic Press, 1980.

[Mars] C.W. Marshall, *Applied Graph Theory.* John Wiley and Sons, 1971.

[Matt2] P. Mattila, *Geometry of sets and measures in euclidean spaces*, to appear, 1993.

[Si-L] L. Simon, *Lectures on Geometric Measure Theory.* Centre for Math. Analysis. Australian Nat. Univ., vol. 33, 1983.

[StWi] J. Stoer and C. Witzgall, *Convexity and optimization in finite dimension*, I, Band 123, Springer Verlag, 1970.

[TiA] A.N. Tikhonov and V.I. Arsenin, *Solutions of Ill-posed problems.* Washington D.C.: Winston 1977.

[Zie] W.P. Ziemer, *Weakly differentiable functions*, Springer Verlag, Berlin, 1989.

III Mathematical analysis of the Mumford-Shah model.

[Amb1] L. Ambrosio, *A compactness theorem for a special class of functions of bounded variation*, Boll. Un. Mat. It., 3-B , 857–881, 1989.

[Amb2] L. Ambrosio, *Existence theory for a new class of variational problems*, Arch. Rational Mech. Anal., 111 , 291–322, 1990.

[Amb3] L. Ambrosio, *Variational problems in SBV*, Acta Applicandae Mathem., 17, 1–40, 1989.

[Amb4] L. Ambrosio-A.Braides, *Functionals defined on partitions in sets of finite perimeter, I: Integral representation and $Gamma$-convergence*, J. Math. Pure Appliquées 69 , 285–306, 1990.

[AmbB] L. Ambrosio and A. Braides, *Functionals defined on partitions in sets of finite perimeter, II: Semicontinuity, relaxation and homogenization*, J. Math. Pure Appliquées 69 , 307–333, 1990.

[AmbTo1] L. Ambrosio and V.M. Tortorelli, *Approximation of functionals depending on jumps by elliptic functionals via Γ-convergence* Comm. Pure Appl. Math., 43, 999–1036, 1991.

[AmbTo] L. Ambrosio and V.M. Tortorelli, *On the approximation of free discontinuity problems* Boll. Un. Mat. It. 3–B 1992.

[BaCG] C. Ballester, V. Caselles and M. Gonzalez, *Affine invariant multiscale segmentation, theory and algorithm.* Preprint. 1994.

[BaGo] C. Ballester and M. Gonzalez, *Affine invariant multiscale segmentation by variational method*, Proceedings of the eighth workshop on image and multidimensional signal processing, IEEE, September 8-10, Cannes. 220–221, 1993.

[BDMP] G. Bellettini, G. Dal Maso and M. Paolini, *Semicontinuity and relaxation properties of a curvature depending functional in 2D.* Preprint, SISSA, Dip. di Matematica, 34014 Trieste, Italy.

[BlatM1] J. Blat and J.M. Morel, *Variational approach in image processing.* Departamento de Matemática Aplicada de la Universidad Complutense de Madrid, 1991.

[BlatM2] J. Blat, and J.M. Morel, *Elliptic problems in image segmentation and their relation to fracture theory.* Proceedings of the Int. Conf. on nonlinear elliptic and parabolic problems, Nancy 88, Longman 1989.

[BPV]   G. Belletini, M. Paolini, C. Verdi, *Numerical minimization of geomet-rical type problems related to calculus of variations.* Pubblicazioni no. 763, Istituto di Analisi Numerica, CNR, Pavia 1990.

[CaLe1] M. Carriero and A. Leaci, S^k-*valued maps minimizing the L^p norm of the gradient with free discontinuities,* Ann. Sc. Norm. Pisa, s.IV 18 , 321–352, 1991.

[CaLe2] M. Carriero, A. Leaci, D. Pallara and E. Pascali, *Euler conditions for a minimum problem with free discontinuity surfaces,* preprint Diparti-mento di Matematica, Lecce, 1988.

[CaLe]   M. Carriero and A. Leaci, *Existence theorem for a Dirichlet problem with free discontinuity set* Nonlinear Analysis TMA, 15 , 661–677, 1990.

[CaLT1] M. Carriero, A. Leaci and F. Tomarelli, *Special Bounded Hessian and elastic - plastic plate,* Rend. Accad. Naz. delle Scienze (dei XL), (109) XV , 223–258, 1992.

[CaLT2] M. Carriero, A. Leaci-F. Tomarelli, *Strong solution for an elastic-plastic plate,* Calc. Var. to appear.

[CaLT]   M. Carriero, A.Leaci and F.Tomarelli, *Plastic free discontinuity and Special Bounded Hessian,* C. R. Acad. Sci. Paris, t. 314, s.I, 595–600, 1992.

[CarDeGL] M. Carriero, E. De Giorgi and A. Leaci, *Existence theorem for a mini-mum problem with free discontinuity set.* Arch. Rat. Mech. Anal., vol. 108, 1989.

[CarLPP] M. Carriero, D. Leaci, E. Pallara and E. Pascali, *Euler conditions for a minimum problem with discontinuity surfaces.* Preprint Universita degli studi di Lecce (Italy), 1988.

[Cham]   A. Chambolle, *Un théorème de Γ-convergence pour la segmentation des signaux.* C.R.Acad.Sci. Paris, 314, S I, 191-196, 1992.

[CoTa1] G.Congedo-I.Tamanini. *On the existence of solutions to a problem in multidimensional segmentation,* Ann. Inst. H. Poincaré, Anal. non lin., 8, 175–195, 1991.

[CoTa]   G.Congedo-I.Tamanini, *Density theorems for local minimizers of area-type functionals,* Rend. Sem. Mat. Univ. Padova, 85 , 217–248, 1991.

[DalMMS1] G. Dal Maso, J.M. Morel and S. Solimini, *Une approche variationnelle en traitement d'images: résultats d'existence et d'approximation.* C.R. Acad. Sci. Paris, t.308, Série I, p. 549-554, 1989.

[DalMMS2] G. Dal Maso, J.M. Morel and S. Solimini, *A variational method in im-age segmentation: existence and approximation results.* Acta Matemat-ica, Vol. 168, 89-151, 1992.

[DaSe3] G. David and S. Semmes, *On the singular set of minimizers of Mumford-Shah functional.* To appear in Journal de Mathématiques Pures et Ap-pliquées.

[DeGi0] E. De Giorgi, *Free Discontinuity Problems in Calculus of Variations,* in:

Frontiers in pure and applied Mathematics, a collection of papers dedicated to J.-L. Lions on the occasion of his $60^t h$ birthday, R. Dautray ed., 55-61, North Holland, 1991.

[DeGi1] E.De Giorgi, *Introduzione ai problemi di discontinuità libera*, in: Symmetry in Nature. A volume in honour of Luigi A. Radicati di Brozolo, I, Scuola Norm. Sup., Pisa , 265– 285, 1989.

[DeGi2] E.De Giorgi, *Problemi con discontinuità libera*, Proc. Int. Symp. "Renato Caccioppoli", Naples, Sept. 20-22 1989, to appear.

[DeGi3] E.De Giorgi, *Free discontinuity problems*, Proc. Israel Math. Ass. Annual Congress 1990, to appear.

[DeGi4] E. De Giorgi, *Nuovi teoremi relativi alle misure $(r - 1)$-dimensionali in uno spazio ad r dimensioni*. Ricerche di Matematica, 4, 1-19, 1955.

[DeGiAm] E.De Giorgi and L.Ambrosio, *Un nuovo tipo di funzionale del Calcolo delle Variazioni*, Atti Accad. Naz. Lincei, s.8 82 , 199–210, 1988.

[DeGiCL] E.De Giorgi, M. Carriero and A. Leaci, *Existence theorem for a minimum problem with free discontinuity set*, Arch. Rational Mech. Anal., 108, 195–218, 1989.

[DeGiCT] E. De Giorgi, G. Congedo and I. Tamanini, *Problemi di regolarità per un nuovo tipo di funzionale del calcolo delle variazioni*, Atti Accad. Naz. Lincei, s.8 82, 673–678, 1988.

[Dib] F. Dibos, *Uniform rectifiability of image segmentations obtained by variational method*. To appear in Journal de Mathematiques Pures et Appliques, 1993.

[DibK] F. Dibos and G. Koepfler, *Propriété de régularité des contours d'une image segmentée*. C.R. Acad. Sci. Paris, t. 313, Série I, p. 573-578, 1991.

[DiSé] F. Dibos and E. Séré, *Approximating Mumford-Shah minimizers*. Preprint, CEREMADE, 1994.

[Kn] J. K. Knowles, *A note on the energy release rate in quasi-static elastic crack propagation*. SIAM J. Appl. Math. 41, 401-412, 1981.

[Lea0] A. Leaci, *Free discontinuity problems with unbounded data: the two dimensional case*, Manuscripta Math. 75 , 429–441, 1992.

[Lea1] A. Leaci, *Free discontinuity problems with unbounded data*, $M^3 AS$, to appear.

[Lég] J.C. Léger, *Une remarque sur la régularité d'une image segmentée*. Journal de Mathématiques Pures et Appliquées, 1993.

[MaTa] U. Massari-I. Tamanini, *Regularity properties of optimal segmentations*, J. Reine Angew. Math., 420 , 61–84, 1991.

[Mod] L. Modica, *Gradient theory of phase transitions with boundary contact energy*. Ann. Inst. Poincaré, Vol. 4 n°5 87, pp. 487-512.

[MorS1] J.M. Morel and S. Solimini, *Segmentation of images by variational methods: a constructive approach*. Rev. Matematica de la Universidad Complutense de Madrid, Vol.1, n° 1,2,3, 169-182, 1988.

[MorS2] J.M. Morel and S. Solimini, *Segmentation of multidimensional data: on the number of regions obtained by "split and merge" methods.* Unpublished.

[MorS3] J.M. Morel and S. Solimini, *Segmentation d'images par méthode variationnelle: une preuve constructive d'existence.* C. R. Acad. Sci. Paris Série I, 1988.

[MorS4] J.M. Morel and S. Solimini, *Density estimates for the boundaries of optimal segmentations.* Unpublished.

[MumS1] D. Mumford and J. Shah, *Boundary detection by minimizing funcionals.* IEEE Conference on Computer Vision and Pattern Recognition, San Francisco 1985.

[MumS2] D. Mumford and J. Shah, *Optimal Approximations by Piecewise Smooth Functions and Associated Variational Problems.* Communications on Pure and Applied Mathematics. vol. XLII No.4, 1989.

[MumS3] D. Mumford and J. Shah, *Boundary detection by minimizing functionals.* Image Understanding, 1988, Ed. S. Ullman and W. Richards.

[Rich1] T.J. Richardson, *Scale independant piecewise smooth segmentation of images via variational methods.* Ph.D Dissertation, Lab. for Information and Decision Systems, MIT, Cambridge MA 02139, February 1990.

[Rich] T.J. Richardson, *Limit theorems for a variational problem arising in Computer Vision.* Annali Scuola Normale Superiore Pisa, 1-49, 1992.

[Sha0] J. Shah, *Properties of energy-minimizing segmentations.* SIAM J. Control and Optimization, 30, 1, 99-111, 1992.

[Sha1] J. Shah, *Segmentation by nonlinear diffusion, II,* Proc. of the IEEE Conf. on Computer Vision and Pattern Recognition, 644-647, 1992.

[Sha2] J. Shah, *Parameter estimation, multiscale representation and algorithms for energy- minimizing segmentations.* IEEE Proc. 10th Int. Conf. on Pattern Recognition (I), 815-819, Atlantic City 1990.

[Sol] S. Solimini, *Notes.* Séminaire de Jacques-Louis Lions, Collège de France, 23 Mars,1990.

[Wan] Y. Wang, *Ph.D dissertation,* Aiken Lab. , Harvard University, 1990

[YuP1] A.L. Yuille and T. Poggio, *Scaling theorems for zero-crossings.* MIT A.I. Memo 730, 1983.

[YuP2] A.L. Yuille and T. Poggio, *Fingerprint theorems for zero-crossings.* MIT, Cambridge, A.I. Memo 730, 1983.

Index

First occurrences and definitions. Numbers without parenthesis means subsection numbers, like 13.2. Numbers between parentheses mean formula, definition or theorem numbers. For every term, we indicate either the subsection where it can be found or the closest formula number situated above.

Segmentation and edge detection algorithms

Notation

(IN APPROXIMATE ORDER OF APPEARANCE)

(x, y) or x, y, z , ... point of the plane $\mathbb{R}^2$

Ω domain of the image

u, g, u_0, u_λ image (a real function defined on Ω

u_x, u_y partial derivatives of u

$Du, \nabla u$ gradient of u

Δu Laplacian of u

K edge set (in general, rectifiable and closed)

$\frac{\partial u}{\partial n}$ derivative of u in the direction normal to K (see 13.1)

$curv$ curvature of a curve or of a level curve of u

div divergence of a vector field

λ, t scale (positive real number)

$E(u, K), E(K)$ Mumford-Shah energy

Card   Cardinality of a finite set

$\mathcal{H}^\alpha$ α-dimensional Hausdorff measure

$osc(u), \omega(u)$ oscillation of a function u

O, O_i regions of a segmentation

$|O|$ two-dimensional Lebesgue measure of a planar set
            (can be also noted $\mathcal{L}_2(O)$)

$\mathcal{L}_n$ n-dimensional Lebesgue measure

∂O boundary of a region

$\partial(O_i, O_j)$ common boundary to two regions

$l(c)$ length of rectifiable curve c

(E, d) metric space with distance d

$d(x, y)$ distance between two points

$d(x, E)$,   distance of a point to a set

$\mathbf{d}(X, Y)$ Hausdorff distance of two sets, 10.1

diam (X) diameter of a set X in a metric space

$\mathcal{A}, \mathcal{B}, \mathcal{C}$ coverings (sets of sets)

$\mathcal{C}(A)$ set of all coverings of A

$|\mathcal{A}|_\alpha$ 6.1

$\rho(\mathcal{A})$ supremum of the diameters of the covering

$\bigcup \mathcal{A}$ union of the covering $\mathcal{A}$

$\mathcal{P}(E)$ set of parts of E

μ outer measure

$\mathcal{M}_\mu$ set of all μ-measurable sets

B^* 2 diam(B)-neighbourhood of B

$\mathcal{B}^*$ set of all B^* for $B \in \mathcal{B}$

$\mathcal{D}_K^\alpha(A)$ $\mathcal{H}^\alpha$-mean density of K on A, 8.1

$\underline{d}_K^\alpha(x)$ lower spherical density of K at x, 8.2

$\overline{d}_K^\alpha(x)$ upper spherical density of K at x, 8.2

$d_K^\alpha(x)$ spherical density of K at x

$K_{m,n}$ (8.9)

$A^{(k)}, A_{(m)}$ reflected sets of A, 9.1

$B(x,r), D(x,r), D_r, B_r, B_r(x)$ balls or disks with radius r, center x

$\mathcal{N}_\varepsilon(A)$ ε-neighbourhood of A

$V, V(x), W, \ldots$ affine subspaces of $\mathbb{R}^n$

P_V orthogonal projection onto V

P_θ orthogonal projection onto the hyperplane orthogonal
    to the direction θ, 12.3

$\int_K f d\mathcal{H}^1$ integral of a function f with respect to the Hausdorff measure
    restricted to K

$C(x, W, \varphi)$ angular neighbourhood of W (9.68)

$C(x, W, \varphi, r)$ angular neighbourhood of V at x with angle φ
    and radius r, 12.1

$C(x, y)$ double cone 12.1

$B(x, y)$ cylinder 12.1

$I(u), I(K), I_B(u)$ terms of the Mumford-Shah energy , 13.1

$\omega_j^s(u)$ oscillation of u (14.22)

Progress in Nonlinear Differential Equations and Their Applications

Editor
Haim Brezis
Département de Mathématiques
Université P. et M. Curie
4, Place Jussieu
75252 Paris Cedex 05
France
and
Department of Mathematics
Rutgers University
New Brunswick, NJ 08903
U.S.A.

Progress in Nonlinear Differential Equations and Their Applications is a book series that lies at the interface of pure and applied mathematics. Many differential equations are motivated by problems arising in such diversified fields as Mechanics, Physics, Differential Geometry, Engineering, Control Theory, Biology, and Economics. This series is open to both the theoretical and applied aspects, hopefully stimulating a fruitful interaction between the two sides. It will publish monographs, polished notes arising from lectures and seminars, graduate level texts, and proceedings of focused and refereed conferences.

We encourage preparation of manuscripts in some form of TeX for delivery in camera-ready copy, which leads to rapid publication, or in electronic form for interfacing with laser printers or typesetters.

Proposals should be sent directly to the editor or to: Birkhäuser Boston, 675 Massachusetts Avenue, Cambridge, MA 02139